Man Eating Bugs

Warning: Although many insects are edible, entomophagy poses some risks. If you are allergic to shrimp, shellfish, dust, or chocolate, never eat an insect. Even the non-allergic, unless in a survival situation, should never eat a raw insect. Certain insects store compounds that make people sick; some are poisonous; others may be carcinogenic. Be as cautious with insects as you would be if you were gathering mushrooms. Know your insects!

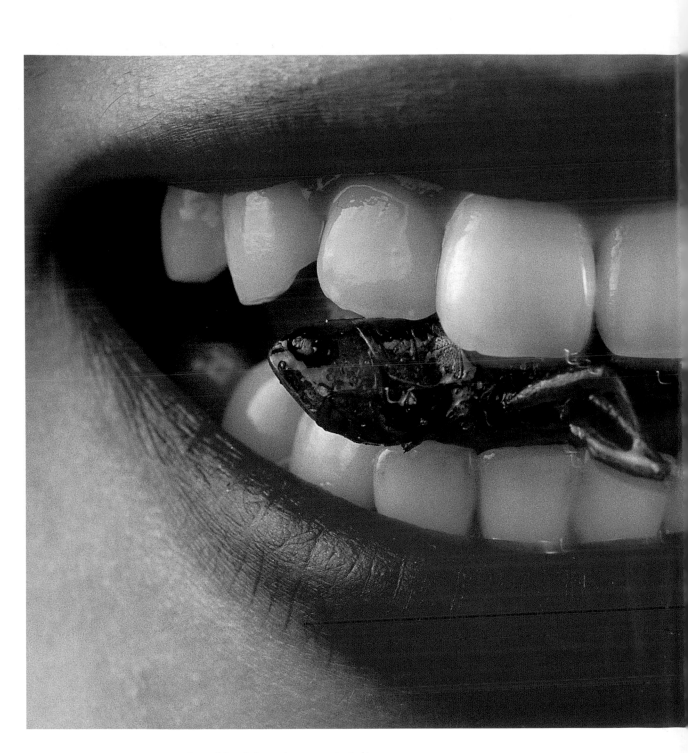

Japan *(above):* A grasshopper marinated in a soy-sugar sauce.

Australia *(following pages):* A live witchetty grub in witchetty grub soup.

Man Eating Bugs

The Art and Science of Eating Insects

Peter Menzel and Faith D'Aluisio

Consulting Entomologist: May R. Berenbaum, Ph.D.
UNIVERSITY OF ILLINOIS AT URBANA-CHAMPAIGN

Editor: Charles C. Mann
Designer: David Griffin
Photographer: Peter Menzel

A Material World Book NAPA, CALIFORNIA
Ten Speed Press BERKELEY, CALIFORNIA

Australia

China

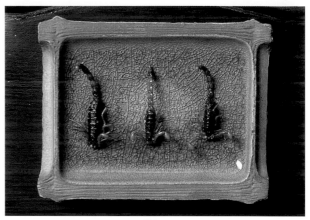

China

Venezuela

Mexico

Cambodia

Japan

South Africa

Mexico

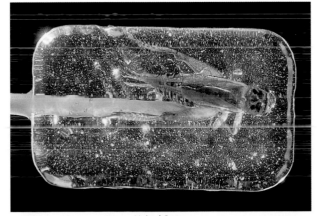

Indonesia

United States

Thailand

Mexico

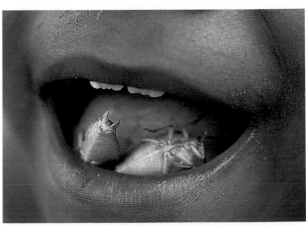

Indonesia

Contents

United States
Mexico
Venezuela
Peru
Japan
China
Thailand
Cambodia
Uganda
Indonesia
Botswana
South Africa
Australia

Australia (*right*): Whipped cream and chocolate topped with a frozen honeypot ant.

Cambodia (*preceding pages*): Deep-fried tarantulas on a skewer.

Foreword
You are what you eat
....

WHEN WE FIRST MET, Peter Menzel did not strike me as a bug sucker, a maggot fancier, a spider chomper, a devourer of ants, a gobbler of grasshoppers, a nibbler of dragonflies and silkworms, or a scorpion scarfer. We were on assignment together in Kuwait, just after the Gulf War. Sabotaged oil wells were still erupting in great gouts of flaming oil, and great blue-black clouds of coalescing oil droplets drifted on the wind, so at times it felt as if we were driving our rental car through a slick purple rain.

I don't recall much of what we ate on that trip. Kuwait City was a ghost town, the electricity only worked sporadically, and we stumbled over rubble in the darkness, with a flashlight, looking for an open restaurant where foreign clean-up workers had established restaurants catering to those of us unfortunate enough to be there. Mostly, I think, we ate noodles. But, after all, it was dark, and after reading this book, I wonder about those "noodles."

I haven't had the pleasure of working with Peter since, though we've kept in touch. A man with a social conscience, he wanted to do a photo essay about the land mines strewn across half the world in the aftermath of various wars. Many innocent people are killed or maimed each year. Too many of them are children.

In Somalia, I heard, he was briefly kidnapped by warlords and had his cameras stolen. It all sounded like typical Peter Menzel stuff.

And then, I began getting bug messages from the man. He was going to the Amazon or equatorial Africa or China, because he had become fascinated with the rather exotic and not entirely explicable concept of insects as comestibles.

During this eight-year bug project, Peter traveled with Faith D'Aluisio, his wife, and her comments in this book, in contrast to Peter's, seem the very voice of reason and epicurean responsibility. The Menzel/D'Aluisio text is amusing and informative. Not only do we learn about worldwide verminous haute cuisine, we can make all kinds of amusing conclusions about the marriage of Peter and Faith.

During the same time period, I published a travel book titled *Pass the Butterworms*. Now, the truth is: There are no such things as butterworms. I just thought they sounded like something you might have to eat, for sustenance, in a distant and culturally unfamiliar area, the sort of place, for instance, where folks honor their dead in above-ground mummification situations (**right**, headman with ancestral mummy in Pummo, Irian Jaya).

In fact, the closest I came to butterworms was a dinner of sago beetle grubs I ate in Irian Jaya, the Indonesian western half of the island of New Guinea. The maggoty-looking delicacies are eaten in the Asmat, the world's largest swamp. My hosts were said to be headhunters and cannibals. I told them, diplomatically I thought, that sago beetle grubs were the best thing I'd ever eaten. In fact, they weren't bad. Lightly sautéed, they had a delicate aroma and tasted rather like creamy snail.

I ate sago grubs out of fear and what I consider to be a kind of noble politeness. Peter Menzel, on the other hand, traveled to Asmat specifically to savor sago beetle grubs. He describes them here as tasting "bacony" and argues, privately, that mine were fried in oil that had been previously used for fish so that my experience with the delicacy was diluted.

Peter says that I haven't really experienced sago grubs until I've eaten them steamed in a sago palm leaf. In fact, the guy is adamant. Every time I talk to him, he bugs me about it.

—TIM CAHILL

Introduction
Where I went wrong: Why I eat bugs

....

I'M NOT A FINICKY EATER—far from it. As a photographer, I travel a lot, but I never carry an emergency food stash. Wherever I'm working, I eat what everyone else eats: camel in Somalia, monkey in Mexico, dog in Indonesia, live squid in Japan, and blood-filled sheep's bladder in Mongolia (only slightly weirder than haggis in Scotland—chopped heart, liver, and lung of a sheep mixed with suet, onions, and oatmeal, then boiled in the sheep's stomach). I enjoyed these dishes not so much because of the taste, but because of the surroundings and the people I was eating with. If you find yourself in the bar scene in *Star Wars*, why order a hamburger?

It wasn't always like that. Back in the fifties, I shared my parents' views that *Wild Kingdom* belonged on TV, not on the dinner table. My mother thought it was adventurous to put a can of Campbell's tomato soup on the meat loaf. I agreed. When I was ten, I was horrified to read in my dad's *True* magazine about Hell's Angels eating live grasshoppers at a motorcycle rally. Eating insects? It was literally the stuff of my nightmares. In one recurring dream, I was finishing a bowl of warm milk and found horrible little living things slithering on the bottom. It woke me up every time.

Thirty years later I had the same feeling (déjà nausea), when I read about *The Food Insects Newsletter* in the *Wall Street Journal*. Founded and edited by Gene DeFoliart, a University of Wisconsin professor of entomology with a wonderful sense of humor, the newsletter concerns itself with the nitty-gritty of entomophagy (eating insects), such as when to raid anthills in Colombia (March to May) and how to cook mealworm quiche or wasp-larvae-and-sour-orange-juice tacos. Here was a dining dimension surpassed only by my nightmares. I subscribed. I ordered all the back issues.

As I pored through the newsletters, a strong force was at work inside me, one that Einstein himself could not have predicted: a simultaneous visceral repulsion and cerebral attraction. I thought bug eating was revolting—and fascinating. My childhood memories egged me on. I had to explore the subject further.

I contacted Julieta Ramos-Elorduy, an entomologist at the University of Mexico City. A middle-aged academic who looked like somebody's aunt (if the aunt were an avid bug eater), she told me about a festival on a mountaintop between Mexico City and Acapulco. In a ritual dating from pre-Aztec times, the townspeople eat *jumiles*, which are traditionally believed to be the souls of their ancestors returning to the living. A *jumil* (*Euchistus taxcoensis*) is a half-inch-long insect—a type of stink bug.

Soon after, Julieta and I were atop Mount Huizteco, near the silver-mining town of Taxco, on the Day of the *Jumil*. Twenty thousand people made their way up the mountain. They drank sodas and beer, listened to music, and hunted for *jumiles* under the leaves of the trees—eating the insects alive or grinding them with chiles and tomatoes into a paste served on tostadas. There was a beauty contest with a blushing teenage *Jumil* Queen (**left**). Here I ate my first insect. Live. It was disgusting. I crunched the unfortunate bug between my teeth before it could scratch my lips or crawl away on my tongue. But my efforts to avoid one unpleasant sensation earned me another: My taste buds were bombed by the creature's bitter, medicinal flavor. *Jumiles*, it turns out, are rich in iodine. Lots of people were watching my reaction (**right**), so I was determined to follow through and swallow. As it went down, I remembered the raw oyster my dad had given me when I was ten. This was worse. At least the oyster wasn't an escape artist with little legs that got stuck in my teeth.

After that first shocking taste, every bug got easier to swallow. In the last eight years I have traveled around the world with my wife, Faith D'Aluisio, a reluctant bug eater, exploring the frontiers of entomophagy. Our view of the culinary potential of invertebrates broadened as we ate raw scorpion in China, roasted grubs in Australia, stir-fried dragonflies in Indonesia, tarantulas on a stick in Cambodia, and live termites in Botswana. Perhaps the most memorable meal was *Theraposa leblondi*, a tarantula big enough to hunt birds, which we ate with Yanomami Indians in the Venezuelan rain forest.

Along the way, a few polite people asked, "Why are you doing this?" (Other, less-polite people said, "Are you crazy?") "It's a way to look at culture, from a very personal angle," I explained. That's true: When we share others' food we close the gaps between us—at least until dessert. But really I should have said, "I want to use entomophagy to encourage us Westerners to examine our own diets and our attitudes toward what we eat."

Every time I returned from a photography trip, I saw my fellow citizens in the supermarket, their shopping carts piled high with packaged food. Although Americans are ever more distant from the food chain, we seem to be ever more shackled to the dinner table. And as the marketplace expands throughout the world, turning the whole globe into Planet Big Mac, so do the waistlines of *Americanus corpus grossus*. The large creatures I see in my supermarket, stuffed with the meat-and-potatoes diet I cherished as a child, look ever more like the future of the human race. Bugs are in no way the solution to this problem—one bite of *jumil* was enough to tell me that—but they are food for thought.

Compared to entomophagy, all the "strange" dishes I had eaten before paled by comparison. Comparison, in fact, was one of the most difficult tasks I had in describing these new tastes to friends. I was able to explain that a witchetty grub cooked in the fire by an Aboriginal grandmother in Australia tastes much like a tender cheese omelet rolled in a smoky phyllo-dough shell. But live termites in Uganda were harder to evoke. After nibbling on several heads (you toss the bodies) I finally figured out the flavor: roasted peanut skins, only juicier. The Ugandan children we were with had no such trouble describing the taste. "It's just like young grasshoppers," they explained.

Our knowledge of cultural cuisine has expanded exponentially; neither my wife nor I having suffered the slightest gastronomic discomfort from these food forays. Insects are a huge, closely related family (the world's largest zoobiomass) and we often wondered if they had a collective memory. Because in the Amazon, by day we ate bugs; by night, they ate us.   —PETER MENZEL

Australia

Surveying the outback, an Aboriginal grandmother searches through the desert for witchetty trees, a type of acacia whose roots may harbor witchetty grubs—the edible larvae of cossid moths. *Inset:* Witchetty grub dip and sautéed grubs; a live grub crawls through the foreground.

Australia

November

PETER: Fifty miles outside Alice Springs. Nearly dead center Australia. You could crawl through the red dirt and scrub for 500 miles in any direction and still be surrounded by red dirt and scrub. I'm following a loaded-down Toyota driven by Bessie Liddle, an Aboriginal artist who has agreed to "go bush" with me for a few days with some of her artist friends. Bessie's carrying a rifle. For kangaroo, she explains. Riding shotgun in my rented Land Cruiser is Kitty Miller, a robust woman with a prizefighter's nose and lips. In the back seat are Pauline Woods and her reticent teenage daughter, Eunice. After a day of digging for bugs in the desert (**top,** Kitty digging), everyone has that special odor that is the exact opposite of the scent cards in magazines. Kitty is pointing out trees and rocks and termite mounds she knows. At times the landscape triggers a memory or matches a vision from oral lore, and she breaks into trancelike chanting as did her ancestors.

Australia is a circle of green around a world of red earth. Landing in Melbourne and flying to Alice Springs is a transition from a clean, fast-moving Establishment city to a dusty, frontier-type place with an indigenous presence that moves to its own rhythm. The landscape is dry, flat, and eroded; Aborigines have lived here for about as long as anybody has lived anywhere. When I contacted Bessie, Kitty, and Pauline, at the Aboriginal Arts Center, and asked them

to hunt witchetty grubs (**opposite,** a handful) with me, I was asking them to do something that their ancestors had done for thousands of years on this same land.

Now and then we veer off the dirt track to chase after a kangaroo, change a flat tire in the shade, or inspect a promising clump of witchetty bushes. Bessie, decked out in a flower-print cotton dress, walks through the bush with an iron digging bar and a shovel. When we see a witchetty bush, she drops her purse in its shade and plomps down to dig. Buzzing flies, shovel scraping through dry sandy soil, then the cracking of a root. Dusty red backlight. She checks if the root has a grub in it. "A little one," she says disappointedly. "If he [were] bigger, might kill more of the tree. But be bigger to eat." In the past, she says, "People could live on these things, along with honey ants and goannas [lizards]. Then the white man showed up. People don't dig anymore. We

still get them when camping out, though. In the bush."

Bessie slips away, hoping to improve her luck. A few hours later, she reappears with a two-foot-long goanna, which she has killed with a shovel. She has a plastic jar of the grubs as well, but she is proudest of her reptilian shovel work. We eat the goanna for dinner after cooking it whole by burying it in the hot sand and ash beneath the fire. It whistles and steams when done and tastes like the lightest, tenderest pork tenderloin you ever ate. Then we turn to the kangaroo tail they bought at a Quickstop on the highway. The women wrap it in tinfoil and toss it onto the fire. I've had delicious kangaroo steaks before, but the tail does nothing for me. It looks like a giant rat tail, cooked in a dumpster fire—a bit too *Road Warrior.*

Finally, the pièce de résis-

tance: the witchetty grubs. Even though this is why I've spent a hot day digging in the bush, I'm skeptical. They are the larvae of cossid moths, after all, not anything I regard as regular food. Pauline tosses the grubs into the fire for a few minutes, then rolls them out of the heat with a stick as her daughter watches (**following pages**). I blow off the dirt and ash and bite down. I'm amazed: The worm's skin is crispy and light; the flesh is creamy and delicate. Witchetty grub tastes like nut-flavored scrambled eggs and mild mozzarella, wrapped in a phyllo dough pastry. Maybe there's a smoky taste, too. It's hard to tell with the smoke from the campfire swirling in my face. No matter: This is capital-D Delicious. Maybe my idea of circling the globe seeking out cultures that eat bugs isn't so crazy after all.

PETER: I didn't find out until Bessie and I were out in the bush that she rarely goes after witchetty grubs any more. The reason is partly that hunting for food in the bush is a nuisance— it's much easier to hit the supermarket—and partly because she's 60 and doesn't feel like doing much trekking any more. She also complains that the grubs are getting scarce. Bessie and Kitty and Pauline have to hunt around quite a bit to find a witchetty bush that is adequately infested. Every stand they inspect seems to be in such good condition that no grubs are present. When the women finally find the grubs, though, they are pleased— a bit of their childhood is returning. Bessie looks absolutely thrilled as she savors a roasted grub (**above**).

I don't expect to see that blissful look again—certainly not in the middle of Alice

Springs—but I do. After returning from the bush and dropping the women at their respective houses (in a typical Alice Springs suburban housing development: cinder-block homes, racially mixed offspring), I clean myself up and visit a restaurant on the Todd Street Mall that has a sign for witchetty soup and crocodile chowder. *La Cafeterie* is owned by Anna and Jean-Pierre Rodot, a French couple who came to Australia five years ago. The affable Jean-Pierre heats up some witchetty soup (which he sells canned) and is amazed and thrilled when I present him with some fresh witchetty grubs, plopping one in his soup. The inevitable joke ensues. "Waiter, there's a..." yeah, yeah. Then (**above right**) he plucks out the grub, licks off the soup, and eats it alive, sucking out the egg-yolklike innards with gusto. His wife, Anna, thinks his gusto is disgusting.

Vic Cherikoff's Witchetty Grub Dip

5 large witchetty grubs
1 teaspoon vegetable oil
1 pinch salt
1 cup low-fat sour cream
½ cup ricotta cheese

Roast or fry grubs in oil until well-browned. Season lightly with salt, then blend to a smooth paste in a food processor with other ingredients. Serve with burrawang bread pieces (burrawang flour has a yeasty cheese flavor and is made from a toxic palm nut after extensive processing) or wattle flatbread (wattle flour is made from the roasted seeds of a type of acacia). (Adapted from *Uniquely Australian*, by Vic Cherikoff.)

FAITH: It's amazing to me that someone came up with the idea of eating a witchetty grub. I tend not to like the taste of fatty foods, and this thing looks like a living, squirming, pasty-white piece of fat, which, of course, it is. But even thinking about this presupposes that I put this grub in the category of "food," which I don't. Or at least I didn't.

When Peter told me we were going to finish the "bug project" he'd been intermittently working on for a couple of years, I was happy, as usual, about the travel involved. But I gave no thought at all to the likelihood that I might have to sample some of our subject matter. Denial, I believe, is the term I'm looking for. Thinking of bugs as food runs counter to my North American cultural heritage. Those of us raised in the home of the brave, land of the Big Gulp and generous-fit jeans—we are a picky lot.

Four continents and a

dozen countries later, I realize that everyone else on the planet is just as picky. It's scarcely an exaggeration to say that you can discover the differences between one culture and another by studying their dinner menus.

I don't know exactly when all this clicked, but I do remember my conversation with David, a Ugandan national policeman who accompanied us into one of his country's many deep forests. It started when he leaned over our container of palm worms, wrinkling his nose in distaste. "How do you know you won't like them?" I asked.

"I just know," he said. His government-issue semiautomatic weapon was slung over his arm; his thumb hooked through the strap.

Fascinating. This was the same discussion I'd had with most North Americans about eating any insect at all. I was struck by a thought and asked him, "Do you eat termites?"

He grinned. "Yes, they are very good."

"Very nutty," I said, having eaten one or two recently.

"I like the grasshoppers, too," he said.

"Well, then, why won't you try these palm grubs if you eat these other things?"

"It is not the same," he said flatly. "I've never eaten these."

David is not alone. In Indonesia, people from the province of Irian Jaya laugh at the Balinese for eating dragonflies. Dragonflies have no meat, scoff the Irian Jayans, who eat cicadas. In China, insect-free citizens of Shanghai turn up their noses at the beetle-eating Cantonese in the south; meanwhile, an hour away by train in the city of Suzhou, people commonly eat silkworm pupae. The Cantonese themselves are split on the desirability of scorpions as food, although everyone

concedes that they are good for medicine. And so on.

People who don't eat insects find the very thought distasteful, although their ancestors might have eaten them at one time or another. Again, it's a matter of taste; why is a fresh raw oyster acceptable as a food to some and not to others?

Why do many Westerners now embrace sushi and sashimi when their parents would have gagged at the notion of eating raw fish? If I know that drawing the line is an arbitrary process, why does the thought of eating certain things still make my stomach turn?

Ask this same question about the delicately flavored sago palm grub in Indonesia or the medicinal-tasting stink bug in Mexico. Or the sugar-coated doughnut.

I gave some children in Thailand a taste of sugar doughnuts from a new Mister Donut franchise. Since the kids had never eaten doughnuts, I thought

I'd see how they liked them. They were extremely unappreciative, preferring their spicy curry and rice.

It is too much of a generalization to say that everyone eats a certain insect, or that no one in a given culture will try a certain dish; there are adventurous souls in every culture. Although I'm inclined to push the envelope in general, Peter had to drag me kicking and screaming into this "bug thing." Eat witchetty grubs? This new foray has forced me to look into the very core of my food prejudices. With every strange object I put into my mouth, I'm finding the exact boundaries of my cultural conditioning and hanging onto them by a thread.

PETER: An hour north of Alice Springs we head down a dirt track parallel to a fence that drifts due east, like a jet contrail, for six miles. Suddenly the women tell me to stop. We crawl through a livestock fence and break branches off a bloodwood tree with dozens of knobby galls (**above**). Bessie calls them "bush coconuts." Later she splits them open with a hatchet. Inside is a chestnut-sized opening, the home of a light green grub: *Cystococcus echiniformis*, a scale insect (**top left**). The female insect burrows under the bark, and her saliva irritates the tree, causing it to form a protective gall around her. The oyster-and-pearl analogy is pretty close. In her dark little

home, the mother wasp hatches several dozen tiny offspring, all male. When the wings of the males mature, Mom then hatches several hundred females, which climb onto their winged brothers, exit through the gall hole, and end up at another part of the tree or in another tree altogether. The big brothers die within a day and the little sisters begin the cycle anew by crawling into the bark to form a new gall. But I don't know any of this at the time—I just suck out the juice from the grub. It has no taste that I can discern. Along with the grub inside the gall are a small handful of tiny nymphs (immature insects), hundreds or thousands of them

(**bottom left**). That they are eaten alive is no problem, since they taste nutty and are so small I can't see them clearly without my reading glasses. Hundreds go down the hatch with every bite.

FAITH: In a way it's foolish to greet the idea of eating insects with revulsion. Certainly one shouldn't think of insects themselves that way: there are too many of them, and they're too important to life. Bigger animals may catch our eye when we walk in forests and grasslands, but it's an invertebrate world. May Berenbaum, an entomologist who helped us learn more about the bugs we

encountered, told us that the estimated total number of insects on Earth is 10 quintillion, or 10,000,000,000,000,000,000. Some termite and ant colonies have several million members; locust swarms can contain up to a billion. Given the preeminence of insects in this world, detesting them all is a little foolish—it's like choosing to hate the house you live in. "We need invertebrates, but they don't need us," entomologist Edward O. Wilson has written. Wilson thinks that if human beings were to disappear tomorrow, the world would not change much. But if invertebrates were to disappear, human beings would be extinct in a

few months, along with fish, amphibians, birds, and mammals. Soon after would go most flowering plants. Within a few decades only bacteria, algae, and a few simple plants would be left.

My friend Cam points out that since most Americans have no connection to insects other than a blanket reaction of dislike, entomophagy at least serves the purpose of getting us to look closely—even if only for a while—at the dominant life-forms of this planet. In the dismayed pause while I decide whether to pop some invertebrate in my mouth, I'm more aware of the insect world than I have ever been before.

Replete with Honey

Honeypot ants (*Melophorus bagoti* and several species of the genus *Camponotus*) owe their name to repletes, specialized workers whose sole function is to store nectar gathered by their nestmates. Too distended with nectar to move, repletes hang upside-down from the ceiling of domed chambers as much as six feet underground. In addition to collecting nectar from flowers, worker ants routinely feed repletes with collected honeydew from plant lice and scale insects (honeydew is the sugar-rich excretory product of these sap-feeding species).

PETER: Bessie and Kitty spot a dead mulga tree, looking bleached against the dark blue sky. They dig down under it a couple of feet. Bingo, sweet success: honey-pot ants. These ants collect sap, digest it, and give it to repletes (**see left**), which store it in their swollen abdominal pouches, trans-parent sacks the size of large peas (**above**). To get my shot, I dig a pit parallel to theirs and lie on my stomach with my head in the hole, ostrich-

glad I am where I am in the food chain.

After going out into the bush, I go to the suburbs of Sydney, where I spend four days with Vic Cherikoff (**preceding pages,** Vic with witchetty grub dip), preparing and photographing insect dishes. An energetic, bearded man with the air of someone who is at home in the woods, he lives in an old adobe house and oversees a staff of four people. Bush Tucker Supply Party Ltd., his company, sells native food to restaurants and stores all over Australia, and on the Internet. Using a screen door on Vic's house, we build an elaborate set to photograph witchetty grub dip, one of his favorite recipes.

In the yard, Vic has built homes for thousands of tiny native stingless bees. As thunderstorms rumble in the distance, we watch them adjust their flight patterns to allow maximum traffic in and out of their nests before the onset of rain. Their honey, called "sugar bag," is delicate and so thin it pours almost like wine (**below**). Vic uses it to add extra sweetness to a special treat he calls "Honey Ant Dreaming" (**page 11**). After filling small chocolate cups one-third full with sugar bag honey, he fills the cups with whipped cream and tops the confection with a frozen honeypot replete.

style, shooting through the chamber. The ants hang from the ceiling of their underground chambers, ready to regurgitate sweet nectar to workers. When the sun shines through them, they focus the light like a lens and look like gems. I have been kidding Bessie about carrying her purse around in the bush, but when she produces a compact mirror from her bag to backlight the repletes, I am grateful.

With my hands around the camera, I am easy prey for the bush flies. More heat and dust and red dirt and flies and sweat. When Bessie finishes harvesting the ants, I am in a killer mood for the sweet slaughter. We sit in the shade with a bucket of ants, picking them up by the head and biting off their swollen little bellies, letting the sweet, warm nectar trickle down our dry throats. I am

Japan

Near Mount Fuji, one of the most important images in traditional Japanese culture, traffic in a construction zone is slowed by a battery-operated robot traffic cop—an image of a newer, technology-obsessed culture. *Inset:* Cans of baby bees and grasshoppers.

Japan

December

PETER: We stop the car to photograph the robot cop on this incredibly expensive toll road near Mt. Fuji (**preceding pages**). Thirty minutes later we're in Ina City, up in the Japanese Alps, drinking green tea with Kazumi Nakamura, a retired fireman who belongs to an elite group of licensed *zaza-mushi* hunters. *Zaza-mushi*—*zaza*, the sound of rushing river water, and *mushi*, insect—are the larvae of aquatic caddis flies (order Trichoptera). Mr. Nakamura graciously provides a plate of them for sampling. I take a bite. All I taste is the sugar and soy sauce used in cooking them. I try a few more of the critters.

Whatever they taste like, it's subtle. Mr. Nakamura sure likes them, though. He nets *zaza-mushi* in the Tenryu River, which slashes through the middle of town, passing under a dozen low traffic bridges. His favorite spot is across from a gravel plant.

Early the next morning we are standing with Mr. Nakamura in the river across from the plant. He's lent us special *zaza-mushi*-stalking waders (**bottom left**). The foot of each boot has a special place for your big toe, Ninja-turtle style. Over these amphibian rubber feet he straps on steel crampons to prevent him from slipping in the *zaza*. To catch the *mushi* he uses a wire net shaped like a small baseball back-

stop. He sets it in the river downstream from where he overturns rocks with his feet and a pickaxe (**above**). Under the rocks live the ugly little larvae of the caddis fly.

After two hours of rocking and roiling, Mr. Nakamura has a bucket full of slithering green pincer-headed little monsters (**top left**). They bear an uncanny resemblance to the creepy scuttling things in the milk at the bot-

tom of my cereal bowl in my recurring childhood nightmares. Back in his tiny kitchen, Mr. Nakamura dumps the live insects into a huge pot of boiling water. Ten minutes later, back outside, he spreads the now-half-cooked *zaza-mushi* over a newspaper on the tailgate of his small truck and cleans away the river debris with chopsticks (**following pages**). Then the insects go back into the hot kettle and are sautéed for a few minutes

with a cup of soy sauce and three-quarters cup sugar for each two pounds of bugs.

When he offers them to me after they have cooled, I can't help thinking about the water of the Tenryu. I'm surprised there is so much trash and plastic in a mountain river. That's not the worst of it. Ina City has no sewage treatment plant, and neither do the towns upstream. In his book *Edible Insects of the World*, Professor J. Mitsuhashi of Tokyo University

states that although *zaza-mushi* are found in almost all Japanese rivers, the ones from the Tenryu River are the best because of the extremely clean water, and that *zaza-mushi* lose their good flavor if they are not harvested in December and January. Hmm. I am glad it is December and very glad that the *zaza-mushi* are cooked thoroughly in boiling water.

Instead of Sushi, *Zaza-mushi*

During winter, a professional collector like Mr. Nakamura can collect five pounds of *zaza-mushi* a day, selling them for about $40 per pound. There are currently about 40 such collectors, many of whom sell to insect canners; canning has been practiced since 1956. Several related insect species are referred to as *zaza-mushi*, but *Stenopsyche griseipennis, Parastenopsyche sauteri,* and *Hydropsycheodes brevilineata* are among the most common targets of collectors. They reproduce twice a year, but the summer brood is not collected because it lacks the tasty fats and carbohydrates the winter brood accumulates to prepare for overwintering.

PETER: This afternoon we visit the Kaneman Company in Ina City—a retail shop selling fancy foods. Tiny by U.S. standards but mid-sized for Japan, the store is apparently the major outlet for insects in town. Everything on the shelves is very neat and very small—just like Mr. and Mrs. Kaneman, serving customers from behind the counter. I bought $40 worth of canned insects: baby bees, silkworm pupae, grasshoppers, and *zaza-mushi*. Canning is the fate of most edible insects harvested in Japan.

Even in a traditional inn in Ina City, insects are served cold, straight out of the can. We take a tiny private dining room and spend five hours shooting insect appetizers: *zaza-mushi*, grasshoppers, bee larvae, and silkworm pupae. Every morsel is exquisitely presented (**above**). But the beautiful arrangements don't disguise

the fact that all the insects have the same flavor and dark brown color, because they are cooked in soy sauce and sugar before canning. None have flavor to tell *Gourmet* magazine about, but I *do* manage to detect a light shrimplike flavor from the *zaza-mushi* beneath the soy. The grasshoppers, bee larvae, and silkworm pupae have almost no taste of their own. Later, in China, I eat fresh, hot silkworm pupae, and their peanut-buttery flavor pops in my mouth. Served cold like this, they beg for beer. Or better yet, hot sake.

Insect appetizers in Japan are not easy to find. Nagano prefecture, which encompasses the mountainous region northwest of Tokyo, including Ina City, is a center for grasshoppers and *zaza-mushi*. This rural area still clings to its peasant culinary traditions; insects are not on many menus, but they are easier to find here than in Tokyo, where only a few

restaurants—either owned by Nagano natives or new establishments fishing for trendy clients—have caught up with the past. Otherwise, most Japanese have the same reaction to entomophagy as Westerners.

I have a chance to learn this firsthand back in Tokyo. In a basement sushi bar (**right**), we meet Mariko Urabe. Her perfect teeth and smile have awarded her the honor of eating insect appetizers for my camera—an honor that she is less than thrilled to win. As we set up a softbox flash, her aversion to edible insects becomes more and more apparent. She thinks we are crazy to eat these things—an odd reversal of the clichéd image of the Westerner shying back from sushi. I order beer in an effort to put her at ease. No luck. I order five small bowls of *inago* (grasshoppers) and eat three of them myself in a further attempt

to make her feel comfortable. That doesn't work either. Still she feels obligated to do it—victimized by the Japanese custom of not being able to say no. She is oh-so-gracious about the whole thing, but she has trouble chewing and doesn't want to swallow the bugs.

FAITH: It's no surprise to me that Mariko Urabe would find eating bugs distasteful—even ones flavored with soy sauce and sugar. Insects have never been a part of her family's diet. Most people's food preferences are established by the age of five or six, and their dietary boundaries expand little thereafter. Although the chemical senses of taste and smell play a major role in whether we find a particular food palatable, the food preferences conditioned into us by our culture are an equally powerful determinant. These preferences can be as simple as the constraints on availability of certain foods or as complex as the deep-seated cultural mores that dictate acceptable and unacceptable foods. Sometimes it's a combination of the two: In the Peruvian rain forest, the palm grub is eaten by the indigenous Machigüenga people but not by the urban colonists who come there to work because they do not consider it food. When Mariko put that bug in her mouth, her tastebuds told her that it tasted almost entirely of soy sauce and sugar, basic elements in Japanese cuisine. But her family upbringing was telling her that the insect was awful. Culture and sensory input warred with each other. As a compromise, she put the bug between her lips—and kept it there.

Thailand & Cambodia

North of Thailand's capital city of Bangkok, at the temple complex of Wat Chae Wattanaram, rows of stone Buddhas—a common image in this overwhelmingly Buddhist nation—testify to his achievement of enlightenment. *Inset:* Deep-fried giant waterbugs with a sculpted tomato garnish in a restaurant in northern Thailand.

Thailand

May

FAITH: Though motorbikes and buses hog the roadways all day long and traffic congestion at times rivals that of Bangkok, the northern Thai city of Chiang Mai is an absolute charmer. It's difficult to describe it without sounding like a guidebook. Perhaps because a profusion of wildly colorful vegetation spills indiscriminately over mansion, house, and shanty alike, what would otherwise be just another junky tourist town is transformed into something more pleasant and alive. The crush of people that would be irritating anywhere else somehow becomes part of the excitement. The night market is not just a market—it's a glittery bauble of a place where vendors sell every piece of clothing imaginable and every kind of food. In this feast for the eyes and stomach, buyers and sellers mingle among the stalls of traditional hill tribe crafts, spices, silks, and hot street snacks. And outside the city is the countryside, hot and green and everything the tropics is supposed to be.

PETER: The stifling, humid afternoon heat has everyone lethargic. I'm barely able to keep awake lying in the shade under the Khuenkaew family's teak house in a rural rice farming village outside Chiang Mai. For the last few days I've been sleeping on the living-room floor on a thin wicker mat and I'm a sore slug. The Khuenkaews are resting, too; they just finished building the new concrete bridge over the drainage ditch to their mother's house. The rice in the field facing the kitchen is

not quite ready for harvest. An old neighbor across the road died a few days ago and the funeral feast preparations are underway. The heat was broken briefly last night by a long series of thunderstorms that heralded the beginning of the rainy season.

I'm here photographing this family for another project, so I'm not thinking about insects except when they skitter around me at night after dropping down from the fluorescent light dangling overhead. Eating them is not on my mind. But this afternoon, after another in a series of heavy showers, some of 9-year-old Visith's marble-playing friends notice a hole near the side of the house from which giant

winged ants periodically emerge and fly off. Word spreads and soon the entire family and a half-dozen neighbors are sitting on the ground or on stools around the bunch of ant holes. It's *maeng man* time, a once-in-a-year event triggered by the first heavy rains.

The ants coming out of the holes are female giant red ants (the *maeng man*), which are the size of a quarter. They have brand-new wings, on which they intend to fly off and start new families. But when they emerge, the Khuenkaews and their friends pluck them up by the wings (**above**). Tiny white worker ants are crawling all over the adults, and the

workers bite like crazy, so you have to either knock the workers off or avoid them by plucking up the giant queens by their wings and quickly dropping them into empty liquor bottles. One nearly full bottle of ants equals two hours of this activity per person working on two or three holes. If you lack the patience or speed for this activity, you can buy the ants in the market (**right,** on banana leaves next to a vendor sorting peanuts).

Later that evening, Buaphet Khuenkaew, mother and family cook, stir-fries the ants. We sit on the floor during dinner with the ants as an extra course in addition to our other spicy vegetables, eggs, and sticky rice. They are delicious; a high fat content gives their crispy bodies a rich, bacony flavor. The Khuenkaews are pleased that I helped them pick the ants and are even more pleased when I show how much I like them.

Early the next morning I am surprised to see the village headman approach the house. He's been standoffish until now, but he makes it clear that he wants to see me. I soon find out why: He's carrying a bowl full of termites that he caught and roasted the night before, after hearing that I had enjoyed the *maeng man* so much. It's a gift for me, he says.

After thanking him, I bring the termites to Buaphet, who adds them to our breakfast buffet. They are not as fatty as the giant ants but quite tasty as well. We eat them with rice, adding new meaning to having a Rice Krispies breakfast.

PETER: A few days after the *maeng man* surprise, I drive with Professor Prachaval Sukumalanand, an entomologist at the University of Chiang Mai, through the frenetic stew of Thai traffic to the Kan Ron Ban Suan Restaurant near the university. I'm relieved to be eating in an open, parklike setting with lots of trees and awnings—a good place to be on this hot tropical evening. This place specializes in insects, which is why we're here.

Prof. Sukumalanand orders icy cold Singha beer and various bug appetizers at our table under a spreading shade tree. The first course is fried bamboo worms (**right**), which are not worms at all, but the larval stage of a moth (lots of edible "worms" aren't really worms, but that's what they're commonly called). They do live in bamboo, though. They don't have a scientific name, since the adults have not yet been collected by scientists. In Thai the larvae are called *rot duan*, "express trains," because they resemble tiny trains. Whatever the name, they are delicious—like salty crispy shrimp puffs. We consume an entire bowl.

Our next course is stir-fried June beetles *(Anomala antigua)*. We eat them by pulling off the wings and legs. Then we suck out the white meat in the abdomen as if they were so many tiny crabs. The bamboo worms and the June bugs turn out to be the highlights. The mole crickets *(Gryllotalpa africana)*, grasshoppers *(Cyrtacanthacris tatarica)*, and giant red ants are okay but not nearly as good as the ones I ate back in the Khuenkaews' village, hot out of the wok.

Later I return to the restaurant and talk to the owner and chef, Bang-orn Tuwanon. She deep-fries some giant water bugs in batter. Each one is as long as one of my fingers. They are caught at night, lured toward a blue fluorescent light that attracts them, then captured with a net. Hoping that the water bugs will be as good as the other insects, I try to bite one in half, but it offers considerable resistance. It's like beef jerky that you don't have patience for. While I work on chewing through the tough exoskeleton, the gooey insides leak into my mouth with a pungent, bug-gutsy taste. I keep chewing on the thing—the sucker is

tough!—but can't manage to get it down. Eventually when I spit it into my napkin, the pungent taste stays on my palate for a while.

I can't imagine people eating this bug. It turns out that the ubiquitous Thai fish sauce *nam pla* is often flavored with the giant water bugs (**left,** in bottles on a Bangkok market shelf). It's called *nam pla mang da.*

Sensitive-Guy Bugs

The giant water bug, *Lethocerus indica*, is a member of the family Belastomatidae, a group of insects with an unusual set of family values. In several species of *Lethocerus*, the responsibility for child rearing rests exclusively with the male. After mating, the female deposits her eggs on the back of her mate and swims away. Because the egg mass on their backs can double their weight, it's harder for the males to reach the surface to replenish their oxygen supply. They are also more prone to attack by predators, including female giant water bugs, which will destroy a brooding male's egg mass and replace it with one of their own. Male *Lethocerus deyrollei* minimize the risk of female egg-mass destruction by brooding their eggs above the surface of the water, where they cannot be detected by roving females.

Bug Eyes

Even though humans cannot see ultraviolet light, insects can—a fact not lost on cricket hunters, bug zapper makers, or plants that depend on insects for pollination. Red, on the other hand, is invisible to most insects. In other words, wearing bright clothing will not attract the attention of mosquitoes and flies, but carrying a plant growlight will.

Cambodia

....

June

PETER: I've traveled to more than fifty countries in my work as a photographer, and Cambodia is way up there on the scale of immediacy of sensory assault. The heat and humidity intensifies the chaotic process of obtaining a visa at the Phnom Penh airport: scramble through crowd for forms, give forms and twenty U.S. dollars (nobody uses Cambodian money if they can help it) to appropriate sweating military bureaucrats, wait while bureaucrats painstakingly write down name and passport number, retrieve luggage from mass of people, push through door, encounter first beggars. A shocking number of people with missing limbs (mostly land mine victims) are begging on the streets. Around these sad intimations of this country's awful recent history roars impossible traffic, low horsepower but high density and high aggression, like bumper cars on crank. Add dust and the odors of exhaust, rickshaw driver sweat, and every imaginable living thing being fried or roasted, to your own adrenal rush every time you successfully cross one of these streets alive, and you begin to get a feeling for Phnom Penh.

When our interpreter's boss learned of our plans to visit the countryside in search of edible insects, she demanded we call the American embassy to check on "security problems," especially for the trip upriver to the ruins of Angkor Wat. An embassy security officer gave us a litany of warnings about travel outside Phnom Penh: "In convoys, if possible, and only between the hours of 9 a.m. and 3 p.m." When I

asked about the trip to Angkor Wat, he said the boat wasn't likely to be attacked, although he knew of recent shootings. He was more concerned about the lack of life preservers on the crowded boats. Several tourists had recently drowned when a boat capsized. The driver of the boat was an 11-year-old boy. Faith and I decided to talk to the locals, who are usually good judges of whatever elements of danger there might be, and we found no shortage of people willing to take us into the countryside. Ultimately though, we decided against taking the boat because the chance for a few good riverside pictures wasn't worth the crowding and heat.

FAITH: Haggling over prices in the chaos of Phnom Penh's wholesale market is complicated by its location, which straddles all sides of a busy four-corner intersection and spills out in all

directions. Handlebars of passing bikes and motorcycles grab at shoppers' sleeves and poke dangerously close to the vats of hot oil full of deep-frying crickets, small frogs, and whole small birds (**above**).

The only option in this fantastic mess is to keep moving or to stand behind the sellers crammed up against the walls of the French colonial architecture. It's 6 a.m. and the market is crowded, but by 11 a.m. the prices will be reduced 60 or 70 percent, the crowds will thin, and the bulk buyers here in the wholesale market will become vendors in the Central Market. We find a lot of insects (**right**, cooked cicadas), but standing in front of them for too long gets us yelled at by the sellers. To appease them we buy deep-fried crickets and fresh ant eggs. Peter tries the crickets immediately and pronounces them delectable.

It's impossible to get a good sense of the market

because of the crowd. The only way out is up. We spy a likely viewing spot on the corner of the main market intersection. Our interpreter, Chung Chotana, called "Tana," leads the way up a narrow set of concrete stairs littered with paper trash and rotting greens. On the fourth floor Tana asks a woman squatting in the stairwell for permission to look from her family's balcony. We smile and nod— the international language. The woman says "no problem," in English. She is scrubbing the heck out of a giant woklike cooking pot and cooking vegetables over a small fire set in the hallway on the floor. The house is narrow and the rooms are filled with sleepy children just waking up and afraid of us. The view from the balcony (**preceding pages**) is awesome: you can hardly tell where traffic ends and market begins.

FAITH: When we can't find any tarantulas in the Central Market in Phnom Penh, we're told to look in Kâmpóng Cham Province two hours north. So we join a spectacular game of road chicken on the highway north of Phnom Penh— bumper to bumper with trucks, scooters, rickety buses and oxen, everyone (except for the oxen) leaning on their horns and zipping around. The sole rule of the road seems to be that there are no rules of the road. Tana tells me that it's not safe to be on this highway at night because of robbery and kidnapping. In this traffic, it doesn't seem all that safe during the day.

As we pass through Skón, a dusty outpost filled with small restaurants and open-faced storefronts, we pass several young women on the roadside, their heads covered by *kromas*, a traditional Khmer cloth that serves as sun protection. They are selling trayfuls of deep-fried tarantulas—two to a skewer for 500 riels ($0.20 U.S.). Pay dirt! Their business is brisk and all the customers are men (**top right**). Why? Twenty-one-year-old Sok Khun (**left**) clears up the mystery. The men think the tarantulas are good for their virility. This is very good for business.

Sok Khun stops talking with us as a military vehicle pulls up and several soldiers jump out to buy tarantulas. This gives me a chance to reflect on an unfortunate circumstance. I've told Peter that I'll eat a tarantula if we find them. Now we have.

PETER: The tarantulas are greasy, but good. The legs are crispy, and each big hairy body is a decent-sized chewy bite that tastes

like...deep-fried tarantula. Faith asks me what they taste like, but in the English language there are no words to describe it. If day-old deep-fried chickens had no bones, had hair instead of feathers, and were the size of a newborn sparrow, they might taste like tarantulas. That's as close as I can come. I tell Faith that she has to try one herself.

FAITH: At this point, Peter has already eaten one tarantula and is getting ready to eat another. I can stall no longer. I break off a leg—it's two inches long, but seems like twelve—and ask if this too is supposed to be eaten. Yes, I'm told, so I do. I'm surprised that it doesn't feel hairy in my mouth because it looks awfully hairy. It does, however, taste oily. I dare Tana, our interpreter, to try one—she never has before—and she does.

Sok Khun and her friends are amused by our joint look of dread at the prospect of our first bite. "This is a very normal food to eat," she says. It doesn't taste bad, but I can't say it tastes good. I'm thinking as I nibble that I've never even *touched* one of these things and here I am *eating* one. Peter makes it very clear to everyone that I'm a lightweight in the Tarantula-Eating Hall of Fame. Big deal.

PETER: I buy half a dozen or so live tarantulas before we leave. I want to use them in a photograph up north at the ruins, so Sok Khun puts them in a small plastic bag, knots the end, and punches out a small breathing hole.

That night at the guesthouse, we leave the bag of tarantulas on the table. I can hear them rustling around inside the bag as I fall asleep. At 2 a.m. I get up to take a leak and notice that the bag isn't moving. I examine it more closely: no tarantulas.

I grab a flashlight from my camera bag and on my hands and knees I rake the floor with light and try to be quiet so as not to wake Faith. Did I mention that I was stark naked? A big hairy spider lurks under a chair. Another is behind a sofa, moving toward the bed where my dear wife is sleeping. I grab them one at a time by their back legs and fling them into the bathtub. After five minutes I have recovered five.

How many were in the bag to start? I can't remember, but after five more minutes spent searching, I find no more. When we get up at 3:30 a.m. to get to the ruins before first light, Faith discovers the spiders in the bathtub and decides not to take a shower. I round them up for their ride in my camera bag to the Bayon temple. We spend two more nights in the room, but we don't find any more spiders.

PETER: Siem Reap is the gateway to the ruins of Angkor Wat. It's bigger than I expect but also much smaller and more pleasant than most towns near internationally famous cultural monuments. Low buildings, no fancy modern hotels, unpaved streets, crumbling Babar the Elephant colonial architecture, garden restaurants without AC. You can walk the main street in the dusk and see crickets flying around the streetlights.

Walking the main drag is, in fact, how we spot a dozen kids gathering crickets. They balance a very long bamboo pole with a bushy tree branch lashed to the end in front of a light, holding it there long enough for several dozen crickets to land on the leaves. After crash-landing the cricket wicket onto the street, the kids pounce on the insects as they crawl out of the leafy rubble. Into a five-gallon bucket they go. The streetlight looks like it's producing a gallon of bugs an hour.

FAITH: In the late afternoon, we meet Mr. Liemh, a retired soldier wearing a sarong, a thin button-down shirt, and a snazzy straw hat frayed handsomely at the brim. He shows us his high-tech cricket-catching system (**left**), which he learned from some Thai people. Two fluorescent blacklights in his yard are suspended high above a clear plastic sheet that glows blue from their reflection. Crickets attracted by the light hit the plastic and slide down its length into a bucket of water, where they drown. Liemh empties the bucket every couple of hours and in the morning the captured crickets are deep-fried (**above**) and sold in the market by his daughter. A small basket goes for 6,000 riels ($2.50 U.S.)—apparently a good price.

"The blacklights are very expensive," Liemh tells us. "We used regular light before, but very few crickets came. I've been using this kind for four years and it works much better." We watch as the lights flicker a bit. "At night," he says, "when everyone is watching television, the power level is low. When they go to bed, the light will be stronger."

Liemh sells some of the crickets to wholesalers, who resell them in Thailand—the border is only eighty miles away. "They want us to refrigerate them to keep them fresh," he says, "but few people here have refrigeration except the people who bring the crickets to Thailand." As he talks, his daughter checks the bucket and empties the crickets that have accumulated into a sack. We pack up our equipment and walk back to our car, suddenly noticing that the ordinary street we walked in daylight is now a shocking blue against the inky sky and is lined with money-making blacklights.

At the sound of the chirp, the temperature is...

There are 2,400 species of leaping cricket, in all of which the male makes a musical chirping sound by rubbing a scraper on one forewing along a row of "teeth" on the opposite forewing. Because crickets are cold-blooded, their activity and metabolism are governed by temperature—increasing with heat, decreasing with cold. As a result, they rub their forewings faster in hot weather than in cold. The relationship between number of chirps and temperature is surprisingly exact. To calculate the air temperature in degrees Fahrenheit from the chirp of the snowy tree cricket (*Oecanthus fultoni*), which is common over most of North America, count the number of chirps per minute (n), subtract 40, divide by 4 and add 50—a formula that can be written as $T = [(n - 40) / 4] + 50$.

FAITH: Our interpreter, Tana, repeatedly warns us that we shouldn't be driving after dark. Though the odds of anything happening are slim, I've learned enough about Tana over the past few days to want to oblige her.

As we explore the markets and the countryside, maneuvering around children in school uniforms and men holding weapons as casually as they would a child's hand, Tana tells me about the breakup of her family during the period the Cambodian people call "Pol Pot time"—named for the man who destroyed the nation in his effort to build the perfect utopian (communist) society.

Tana: "We weren't afraid those first days. The Khmer Rouge were welcomed into the city as heroes after they toppled the [U.S.-backed] government of Lon Nol in 1975. They told us they were going to rebuild our city, which had been destroyed by bombs. Every person in Phnom Penh walked into the countryside as they told us to. We expected to return after the reconstruction." Instead, the new regime forced the people of Cambodia to stay in the fields, live communally, and work for the state during a genocidal reign of terror.

"My father was a university professor. In the early days after everyone went into the countryside, he was asked to come back to Phnom Penh with the other intelligent people in the country to help the new government rebuild. My mother asked to go with him but was refused. [Later] we found out that everyone who went back was tortured and killed. People wearing eyeglasses were killed because they were thought to be intelligent. We never found out what happened to my father.

"Everyone worked in the fields to grow rice for the government. My job was to go with the cows. We never had enough to eat so we would catch rats in the field and eat them. They were very delicious.

"The Khmer Rouge destroyed our monetary system. My mother had a [wealthy] friend who never understood that money wasn't worth anything anymore. I could never make her understand. She kept trying to give it to me and others—mostly children—in exchange for favors and food. She finally died."

By 1979, when the Vietnamese toppled the

regime, the Khmer Rouge had killed more than a million Cambodians; a million more had died because of the harsh conditions.

Tana stayed with her mother's family in Kâmpóng Cham province while waiting to return to the city. When Phnom Penh was reopened, Tana's mother and three other women walked for two days straight to get home. Tana rejoined her mother and sister but the loss of her father had destroyed their home. "We are not a good and happy family because there are so few of us left," she said, "and whenever there is a fight or a bomb goes off now, we worry about each other. We want to be together if anything happens again."

Danger continues to lurk in this country long after the fall from power of the Khmer Rouge and the death of Pol Pot. The instability of

the government and its ready use of the military to quell the opposition is as unsettling as the legacy of unrecovered land mines and the spreading proliferation of highway bandits.

Though the Cambodians we meet are wonderfully kind—a family outside the famous ruins of Angkor Wat goes out of their way to cook us the weaver ants (*Oecophylla*) they shook from trees (**left,** stir-fried) and dig up tarantulas for us—we are sobered by the young gun-toting soldiers we see everywhere.

One measure of the great beauty of the country, I suppose, is that I remember the ruins (**preceding pages,** boy fishes near Angkor Wat) even more than the weaponry around them. On the days we set up for photo-graphs at Bayon Temple (**above,** crickets and tarantulas), two miles from the center of the temple complex, the general message is "don't be there before dawn or after dusk." When we linger too long one afternoon, trying to catch the last bit of daylight, our guide murmurs quietly into his radio and a barefoot adolescent in a sarong appears, carrying a semi-automatic weapon. The teenager stands guard silently in the twilight as we pack up our equipment.

Indonesia

The world's fourth most populous country, Indonesia is home to some of the world's most beautiful rice terraces, including these on the island of Bali. *Inset:* Dragonflies, dewinged, salted, and fried in coconut oil, with sweet pepper garnish.

Bali, Indonesia

April

FAITH: You can easily imagine what Bali must have been like before it was overrun by tourists. It is still an island, literally, of Buddhist culture in one of the biggest Muslim countries in the world. But much has had to change to make room for the tourist business that drives the local economy, even though that business is based on keeping as much as possible of the island's traditional way of life.

Many of the rice paddies that surrounded the village of Ubud, on the lush southern volcanic slopes, have long since been replaced by shops full of postcards, wooden carvings, and cheap sarongs. But local nostalgia is tempered by realistic appreciation of economic growth. When we asked our guide, Darta, if he liked Ubud better as it was before, he remarked only that it was different, and today

there is enough money. "Where before it was a struggle to fill the bellies of the children," he says, "now it is not." He remembers his mother setting out, kids in tow, to catch dragonflies to feed their hungry family.

PETER: Although Darta had hunted dragonflies as a child, he didn't ask his kids to come with us because he knew they would refuse. "TV and chicken killed the dragonfly," he said, explaining

that kids today (especially his kids) were lazy, watching too much TV and eating chicken. Hardly anyone catches dragonflies now. Darta himself didn't have much heart for the chase. It was too hot. After an hour or so, he had five dragonflies.

FAITH: After many inquiries we meet Iwayan Darsana, 41, an artist (and part-time tourist guide, as are many

people on the island) who remembers dragonfly hunting fondly. Although chicken replaced dragonflies on his dinner table years ago, Darsana taught his children how to hunt the insect using a slender strip of palmwood dipped in the sticky white sap of the jackfruit tree. When the insect touches the strip, it sticks fast to the sap.

With Darsana in the lead, we go into the fields. Homemade wands whip through the air in the hands of the six excited children. In single file the kids navigate the mushy narrow walkways that separate the rice paddies, brows furrowed in concentration. It's hot and there's no shade. At first it's hard to see any dragonflies, but the kids fan out through the fields nonetheless.

Standing in one paddy, Darsana shouts encouragement as his 8-year-old daughter, Ni Wayan Sriyani, slowly extends her bamboo pole as far as she can reach (**top left**). A dragonfly approaches, zig-zagging over the rice. Like an expert fly-fisher, she flicks out the end of her pole and catches the wing of the first dragonfly of the day. Darsana claps his hands in excitement. Perspiration drips down the length of Sriyani's face, but she smiles brightly as she skewers the dragonfly onto a thin piece of palm.

Spurred by the thrill of the capture, the other children find dragonflies, threading them on the palm strips like dangly, glittering beads on a string (**preceding pages**). Next is a short swim at the waterfall (**right**) where Darsana used to play as a child. Then the family returns home to fry the cache of dragonflies in coconut oil (**bottom left**) and pop them in their mouths like candy.

PETER: After our great dragonfly hunt, we went to the Ubud market at dawn (**above**)—a visual treat. Behind the rows of streetfront boutiques selling everything offered at The Gap for half the price are hundreds of stalls with lotus pods, rambutan fruits, lychee nuts, edible cactus pears—you name it. Flowers are everywhere, especially marigolds, which are used for offerings at Buddhist temples.

FAITH: When Darta finds bee larvae at the market, he tells us to buy it right away—the stall is almost sold out. We follow him to where a woman stands with a tray of the larvae (**right**). "Everyone wants this, but it's expensive," he says. I peer over Darta's shoulder to see what the fuss is about. I know it's honeycomb, so I'm vaguely expecting something honey-colored. Instead it's a dingy

brown color and looks like its been run through a wash cycle a couple of times with the dark clothes. How do you eat this stuff? I ask, and Darta tells me the comb is boiled to get the larvae out. Then the larvae are cooked with coconut oil, garlic, onion, chiles, lemon, fermented fish, sliced green papaya, long beans, and greens.

Faith: And the beeswax is discarded?

Darta: The beeswax from the honeycomb is put in the

The next week we went to Irian Jaya, which is the western, Indonesian half of the island of New Guinea. Our first meal there unexpectedly turned into a birthday barbecue for Bob Pelege, who later helped us get into the interior of the country. Instead of potato salad and hot dogs, the featured dishes were sweetened sticky rice steamed in bamboo, and dog. Putting on the ritz meant putting down the pooch. Bob's friends and family collared two strays, put one into a sack, and bludgeoned it to death. The other escaped, and Bob was apologetic that the huge wok of chopped dog and spicy peppers was not chock full. The warmth and hospitality with which we were being treated suggested that an outraged refusal to partake would be bad form. Thinking of our own 14-year-old mutt at home gave me a twinge of guilt, but I politely ate—two servings. Faith, horrified, took a small bite to be polite and then surreptitiously dumped her portion onto my plate when I wasn't looking. I wouldn't purposefully seek out dog for food, but I must admit it was good.

Mister Twister

Dragonflies are known for their peculiar mating habits. As is usually the case with insects, the male's genital opening is by the tip of the tail, on the ninth abdominal segment. Its copulatory organs, however, are on the second abdominal segment, much closer to the head. Prior to mating, the male moves sperm from the genitalia to a storage sack just behind his legs by looping his abdomen up and around. Once he finds a female, either on the wing or resting on a plant, he grabs her with the structures at the tip of his tail and secures her behind her head; she then has to bend her body up and place her genital opening, at the tip of her abdomen, up against the storage sack. Even after sperm has been transferred, males will sometimes retain their grasp, forcing females to lay eggs under their supervision.

hair of Balinese teenagers during the tooth-filing ceremony [during which the] top teeth are filed down. Balinese believe we have six bad things in our body—lust, greed, jealousy, wantonness, anger, and confusion. When we are teenagers, we are very emotional and aggressive. So these must be controlled, and this is how we do it. Filing is a symbol of throwing out the bad spirits in our body and beeswax is part [of that ceremony].

PETER: At the market and many other places in Indonesia we saw scraggly dogs dozing in the sun; one Ubud market dog had an incredible skin disease and resembled a giant pink naked mole rat. The general lack of reverence for dogs (puppies are petted but adult dogs are eaten) led us to an awkward moment.

Irian Jaya, Indonesia

March

FAITH: Bali and Irian Jaya are both in Indonesia, but it takes a series of island-hopping flights and a full day to get from one to the other. Irian Jaya—the sparsely populated western half of the island of New Guinea—has a contentious relationship with both Indonesia (it chafes under the government's rule) and Papua New Guinea, the nation that occupies the other half of the island.

From Bali to Jayapura, Irian Jaya's northcoast capital, we fly to Timika on the southwestern coast. We then charter a small boat for a seven-hour trip south on the Arafura Sea. We're headed for the Asmat, the swampy southern region, to look for sago grubs. This jungle area is home to hunter-gatherers of Melanesian descent who live generally as their ancestors did. For now, anyway. Some blame the decaying of the Asmat culture on the religious missions that arrived in the 1950s, others on the Indonesian government for letting big timber companies operate here. Whatever the cause, given the way this world works, "progress" is inevitable, and people here are beginning to see much of it. The Asmattans living near the settlements seem pragmatic about the changes. When I ask these villagers why they converted to Catholicism and not to, say, Islam or Protestantism,

the common response is: They got here first.

PETER: *Field notes, 26 March.* Left the Asmat's principal town of Agats (jungle town built on stilts) at 7:30 a.m. in a 40-foot, green, wooden longboat with 40-horsepower outboard. Some drizzle, then hard rain. Over three hours up the river, huddled under a tarp. Motor up to Komor, a thatched village on the muddy Bo River—high dock, dozens of kids (**preceding pages**). The big tides affect the rivers here—the docks are on 15-foot stilts. Clamber up crude ladder with equipment. Nothing lost to the river. Then to the longhouse, the men's house (called a *jeu*), facing the river. Inside the 80-foot-long bamboo-thatch structure on stilts are lots of sleeping men and seven smoldering fires. Sit down, meet assistant

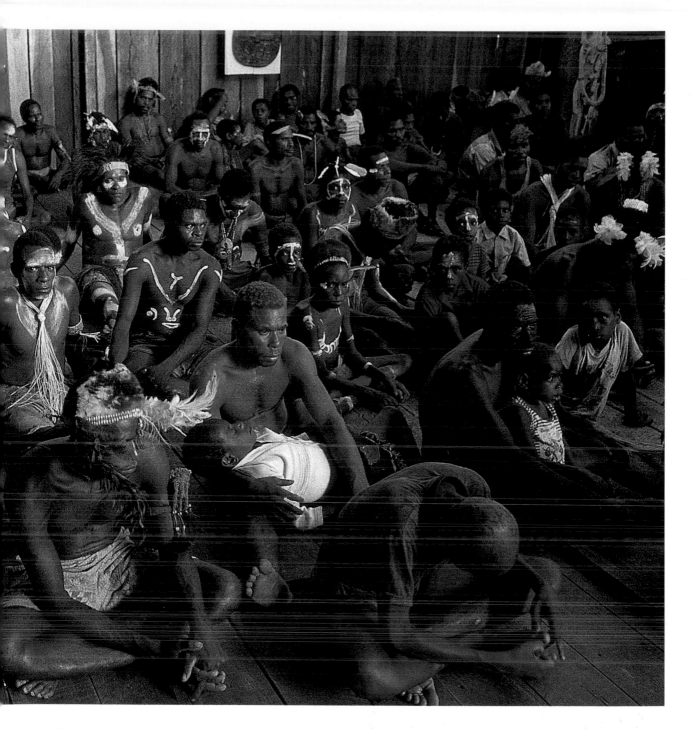

chief and eighty men. Dozens of diseases present. One guy with testicles as big as footballs. One withered arm. Several nearly blind. Skin diseases to fill a textbook. All the men are smoking. Hot, hot. We sleep in our tent in men's house. Like a sauna in a TB ward. Coughing, hacking, farting, talking all night. Sleep naked, drenched in sweat. Up at 5:45 for misty first-light photos of people getting into their muddy dugouts on the chocolate-colored river. Good Friday. Sweltering day shooting people gathering sago grubs (**left**) from interior of sago palm. Back in time to catch Brother Jim's Mass in the village church (**above**). Very hot. Men and women enter separately through doors on either side of the church. Catholic/native service— smells and bells, feathers and paint. Huge hunks of heart of palm carried in as offerings by men and passed around during the service. Sharing. Missionaries rise a notch in my book. After sunset we shoot a family cooking the sago grubs. About eighty people pack into the house to watch us shoot. A fight starts between two men over a woman, and people take sides and rush out of their houses carrying axes, clubs, and bows and arrows. Lots of shouting and screaming but it is settled without bloodshed. Flashlight dies. In the pitch dark, balancing on a rickety board-walk while loaded with equipment, I have to stop like a blind man crossing a dangerous street until some kids come by with grass torches to help. At 10:45 p.m. we move the tent outside the longhouse where the air is less stifling. A heavy rain starts at 11 p.m. I am so hot I stand in the pouring rain and rinse off.

FAITH: Thursday—or is it Wednesday? We didn't get much sleep last night. Peter set up our tent with great difficulty surrounded by a huge audience. It was very funny. Our crew helped, which made things worse. Ronny, the porter, found an extra piece of pole and pretended to hit Luki, another crew member. Luki fell over in mock death. That was the high point of the evening. Earlier I had a good conversation with Theo, the assistant chief, in the presence of the men, who seemed to have nothing better to do than roll cigarettes and hack between puffs. I thanked them for letting us visit. Theo leads a thank-you voiced by the group as one low beautiful note and a crescendo-like shout: Waah!

PETER: *Field notes, 27 March:* We take Plipus Manmank's family (three couples and one small boy) in our longboat from Komor village up a tributary of the river Bo. The river turns into little more than a stream and gets more and more narrow over the couple of hours we travel. Very shallow, narrow. Logs, mud, maneuvering. Walk into jungle for a half-mile. Find rotting sago-palm tree, cut it down with axes—hack, hack, hack, whomp!—the swampy ground reverberates like Jello. Find some grubs right away. Hot. Everyone's hands are covered with the slimy fermented pulp of this rotting palm as they reach in to grab the grubs. Flies, mosquitos. Grubs are put on a spit and barbecued. Eaten by kid and his dad (**upper and lower right**). Good-enough taste, fatty and bacon-flavored. Very chewy skin however.

Sago processing (**above**) is hard sweaty work. After downing a big sago palm, they hack away the hard fibrous brown exterior of the trunk, exposing a softer fibrous white interior. This is pulverized with a special tool like a blunt stone axe. A sluice is constructed out of palm casings and leaves and the mashed pith is mixed with water from a 3-foot hole dug in the forest floor, which quickly fills with water. The finely pulverized particles are washed out of the fibrous mash and for a couple of hours settle and solidify in the lower part of the trough. The women cut the starchy white rubbery mass into blocks, wrap the blocks in leaves, and carry them out in string bags. One block is covered with dry leaves and torched. This toasts the outside, and after the ash is brushed off it is eaten. Like chewy smoky pudding skin.

FAITH: Excerpts follow from a long riotous chat we have in Sawa village (two hours by motorized longboat from Komor) with Roman Catholic priests Vince Cole and Virgil Peterman, from the U.S., who have been missionaries here since 1979.

F: Have you ever eaten sago grubs?

Fr. Vince: Oh yeah. I eat them a few times a year maybe.

F: So it's not something that repulses you?

Fr. Vince: I wouldn't order them in a restaurant.

Fr. Virgil: Oh I would! I'd like to see what they'd do with 'em in a restaurant.

[General laughter while we wait for Father Virgil to calm down]

F: The sago grub is an important food source here, isn't it?

Fr. Vince: I suspect it's one of the few sources of fat [in the Asmattan diet]. Most of the other foods they eat have no fat. It's not like pork at all—have you ever seen wild pig? It's red like beef. Real lean beef. Anthropologically, the people here look like the anatomy charts in a classroom—they have very little subcutaneous fat. But what

I've noticed is that they don't have good endurance. If they [eat] more food than I do, they can beat me every time, but I noticed when I was on a long trip upriver with them, and we were running out of food and had to ration ourselves for the last couple of days, I had more endurance because I have more body fat in reserve.

F: When did rice become a more regular food to eat?

Fr. Vince: Maybe five or eight years ago. Maybe even less than that. In the more remote villages it's not that common, but here, where there are stores that now sell the rice, it's more common. [Asmattans] don't eat meals on a regular basis. They eat as they are hungry.

Fr. Virgil: Over the last year or two, people have been harvesting and selling the [aromatic, valuable] gaharu wood, and so they have money they didn't have before and they're buying the rice. But actually, it's more of a feast food that they don't ordinarily eat. Like the sago grub.

Fr. Vince: I see people buying rice every day in this village.

Father Virgil swallows a bug of unknown origin and the interview stops while he gags. "It was not a good one," he points out. When he's able to talk, I continue:

F: I hear that you like to eat the sago grub, Father Virgil. How would you describe its flavor?

Fr. Virgil: When I first tried it, I thought it had a flavor similar to bacon. A friend's wife had a way of frying these sago grubs so they became crispy. Real light. I could never do it—they would just burn in the pan—but she would spin 'em around and around 'til they were just shells. They were very light. Another way is to steam them. You fold them into a long leaf and just lay them that way in the pan. When it's opened, you have steamed grubs. They're very sweet. Some are sweeter than others.

FAITH: At daybreak, men, women, and children in the villages of the Asmat bail rainwater from their dugouts and pick up their paddles. They step gracefully into boats—narrow dugouts that, amazingly, don't rock a bit. They lay a piece of smoldering log in the bottom for lighting hand-rolled tobacco-leaf cigarettes throughout the day. The river water is colored brown from mud, silt, and human waste. One or two families go fishing, but most carry axes, machetes, and wooden thrashing tools to harvest sago starch in the forests. Sawa villagers, standing upright in their boats, paddle downriver to family-owned land, faint wisps of smoke trailing behind through the hazy morning light.

It's enlightening for anyone who lives in a developed country to spend time among people who live off the land. Where I see giant palm trees, hanging vines, and lush undergrowth, Asmattans see an age-old supermarket. This morning we go along with Rufina Dochan, her husband Victor Aunum, and her sister Udelia Toronam on a shopping trip, pushing off the muddy bank and heading north on the Pomats River. Even at daybreak we can tell the weather is going to work itself into another stunningly hot day.

The shoreline is both

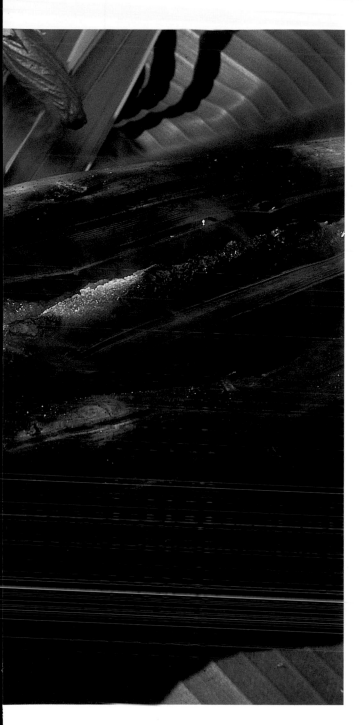

the sago and a good-sized pile of sago grubs—maybe eighty in all —which Rufina and Udelia are going to cook.

Back in the village the two women prepare something like sago grub tamales (**opposite**). Rufina says the dish has no name: It's only sago grubs and sago flour wrapped in sago palm leaves roasted and steamed in the fire. A big audience of intensely interested family members watches us watch the two women work (**preceding pages**). The women press fresh sago starch—it looks like damp grainy flour—into a palm leaf that is sewn into a long bowl with the sharp spine of a palm frond. Then they crush a grub and spread it onto the starch. Its oil makes the starch more gooey, like a pastry crust. Then they arrange the rest of the grubs on top of the crust, fold the edges of the leaf and the "pastry" around the grub mixture, and sew the package shut with palm frond spines. There are no spices, they tell me, and no other ingredients. It's just simple food, says Marius, our translator. Simple indeed—the Asmat diet is so incredibly restricted that people here don't use salt.

The food bakes at the edge of the fire for half an hour until the outside is a bit charred and smoky. Then Rufina pulls out the palm spines and peels the leaves away, revealing a cooked pastry that I'm repulsed by because I see the grubs, but appreciate despite myself because the dish looks good. When we taste it, the crust is chewy and sweet and the grubs taste like fishy bacon. The flavor is good, though I'm careful to take only one sago grub in my portion. It is, after all, a grub.

Grub Work

In addition to providing variety in their diet, sago grubs help Irian Jayans use their time efficiently. When women cut down sago palms to make sago flour, they leave behind the stump and the section just below the crown, both of which have relatively little of the starchy pith used for the flour. The leavings are soon invaded by *Rhynchophorus ferrugineus*—that is, sago grubs—which the women harvest in due course. Meanwhile, men hunt grubs in yapay sago, another variety of sago palm, which is known to be low in pith. In both cases the use of grubs exploits the tree as a food source while avoiding the work of processing large amounts of low-yield sago pith. Indeed, *Rhynchophorus ferrugineus* chews through pith with such efficiency that there have been some tentative stabs at creating commercial grub ranches in the forest. Unfortunately, the farms tend to attract wild pigs, which regard sago grubs as a principal food source. Eating the pigs attracted to the grub ranches would not be an ideal solution because sago grubs are lower in fat and higher in calcium and riboflavin than pork.

algae-green and mud-brown and crammed with a sinewy jumble of mangrove trees, all giving their roots a good soak. The river itself is an obstacle course of floating logs chopped down by locals for the new and burgeoning timber industry. Our pilot deftly maneuvers around them as far as he can along this rapidly waning tidal tributary, and then we walk through varying depths of murky water and watery glop to the family land. We follow

Rufina and her husband into the jungle, tramping through thick mud and heavy mats of rotting palm fronds and around the logs blocking the trail. Along the way the jungle canopy shields us from the direct sun, though the places where it pierces through are blinding. We slog ahead as fast as the shin-deep mud will allow, but the effort of forward motion pales in comparison to the chore of harvesting the sago. Hours later we return with

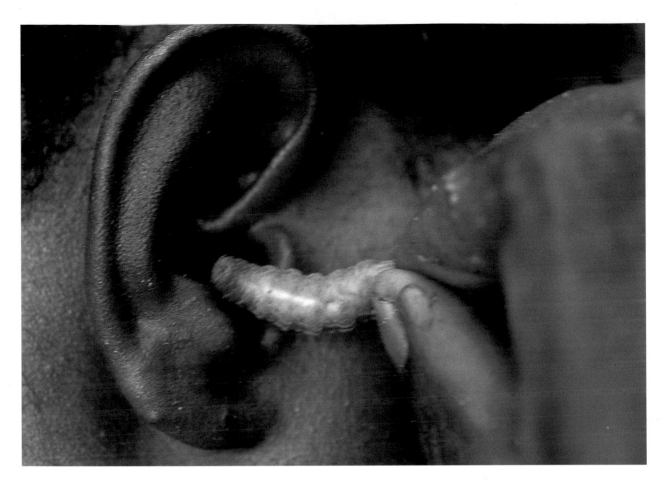

PETER: *Field notes, 30 March.* Jungle logging camp down-river from Sawa village—right out of *Apocalypse Now* with a shrine for the trees killed here. Supposed to atone for forest harvesting

Debugging

Insects have a way of ending up in people's ears. In one study, 27 of 134 foreign objects found in children's ears were arthropods—21 cockroaches, an ant, a fly, three spiders, and a tick. Doctors remove arthropods from auditory canals by vacuuming out the creatures, dislodging them with mineral oil, immobilizing them with a topical anesthetic, and shining a flashlight to lure species attracted by light. *The New England Journal of Medicine* once published a study that compared the efficacy of extraction methods when a patient turned up with a cockroach in each ear.

beyond their needs (they're selling the wood, not using it). Logging by hand is hard, dangerous work. Everyone is friendly, if not too healthy.

I ask if it's true that people use worms to clean out their ears. [I'd heard the story that the worms eat ear wax.] Immediate demonstration with a very small grub (**above**). "You have to hold onto the tail," they caution. "Never let go. You don't want the worm getting lost and coming out the other ear." Are they pulling my leg or putting a bug in my ear?

Off the path we find a jungle chicken nest, like a 6-foot-high compost hump. Digging with their hands, some women turn up an ostrich-sized pink/brown egg, highly prized. *Ulat-kayu* is next—another grub. An axe-man chops open an old log—not a palm but a hardwood. Finds some in tunnels made by eating their way through the log, just under the bark. They're eaten live. I shoot kids fooling around

with grubs (**top right**) and then eat one for solidarity. Juicy. Woody taste to juice. But skin is very hard to swallow—like old chewing gum. We also find tiny bees: not much sting or honey. Frogs, too. I persuade Rufus to put them on his shoulders (**above**). Frog epaulettes.

Logging methodology: Huge hardwood logs, all cut by hand axes, are notched like crochet needles on one end

so that a long vine can be securely attached. Then a dozen men pull and lever this many-ton monolith over skids made of smaller logs (like railroad cross-ties) on the swamp floor for miles, shoving the logs to the river, where they lash them together and float them down to traders who buy them either with money or outboard motors. This is changing the landscape for miles inland. Sometimes you can't see the forest for the Evinrudes.

FAITH: A government logging initiative is putting cash in the pockets of Asmattans who are new to the idea of a monetary system. The money often goes to buy foods new to these areas: instant noodles (**left**), white rice, and the rounded lumps of fried dough called *oli* balls. "They can't fry them fast enough," Father Vince told us. Traditional foods are being shunted aside for these new convenience foods. Irian Jayans have always suffered from vitamin deficiencies, but, as Father Vince says, "It hardly seems a cure to ply them with processed foods." Lucky for Asmattans, the sago palm has no timber value so its flour and grubs are in no immediate danger. In twenty years, I wonder if they'll still be harvesting it.

PETER: *Field notes, 30 March.* We don't leave Sawa until it is dark. Raining again. No moon on the river as we head downstream. Starry holes in the clouds overhead. Lightning in the distance flashes on the river banks. There is a phosphorescence around the longboat in the river. In such a dark night it appears surreally bright. Lightning bug colonies blanket the riverbanks every few hundred meters. They swim in the air in a slowly rotating cloud. Their lights are steady—they don't blink on and off—and the huge glowing coils are reflected in the water. Bioluminescence in the black water and the insects—one of the most amazing displays of light I've ever seen. That we are motoring down a river in a leaky wooden boat, a thousand miles from nowhere, only makes it more intense. Equally amazing is the place we go next, in the highlands of the Baliem Valley, where outside the movie theater in the town of Wamena a drama of Tribal vs. Western Culture is playing continuously (**following pages**).

FAITH: After the Asmat jungles of Irian Jaya—a muddy, sweaty swamp of a place—the cooler Baliem Valley in the highlands is a relief. The men's native dress here is a bit of a surprise (**preceding pages,** penis-gourd clothed man with others in front of the Wamena movie theater). Soon after coming to Soroba village, we're following its children on forest trails. They're hunting stink bugs to eat as a mid-morning snack. In this fertile farming valley inhabited by the Dani, gardener-warriors and former headhunters, the staple foods are taro and yams. Insects are a special protein treat for the younger generation, though adults admit a fondness for the meaty cicada. The older boys leap up tender saplings with hands that cup the trees as they climb (**top left**).

Standing below, I catch a whiff of an unpleasant odor: stink bugs. The boys wrinkle their noses but press on. When their prey is within reach, the children softly call out "mo, mo, mo, mo,"—it keeps the bugs still, they say.

Down below, the younger kids hunt another treat—*mulikaks* (spiders). The boys twist sticks into the large webs, and the spiders are caught in their own nets.

The older kids hand down bugs to the younger ones and they fashion "bug packages," leaves wrapped around a fistful of stink bugs (**following pages**). In an hour the children have enough stink bugs. About twenty makes a good snack for one child. Placing their packages on the edge of a wood fire (**below left**), the children roast the bugs. The crisp shelled insects are eaten whole (**right**). They taste better than some worms I've tried.

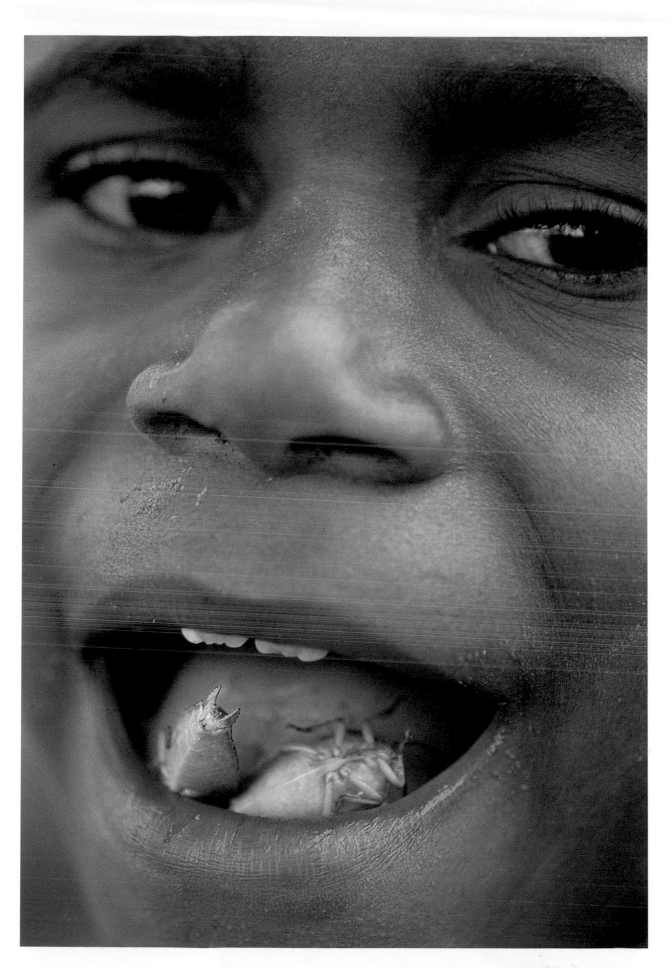

FAITH: Amuloke Walelo is preparing today's vegetables with her husband, Siba Himan, the chief of Soroba village in the Baliem Valley (**right**). "I'm hungry," he says irritably. "Hurry up." Am I imagining it or is Amuloke grinning behind his back? Siba squeezes *buah merah* (a red fruit) onto the taro and potatoes. The blood-red juice makes the vegetarian feast resemble a fresh animal carcass. I'm startled to see that the four fingers of Amuloke's left hand are stumps without fingertips. Her fingers were severed at the first knuckle when she was five years old. In the Dani culture, women lose their fingertips as a tribute to family members who die (**below,** a woman's hand). Not men, though. I ask her about the difference between women and men.

Faith: I've been told that women's lives are so hard here that some kill themselves and their children to escape.

Amuloke: The husbands are angry every day and they make big problems our whole life. They fight and say bad things and are jealous. We feel very sad and feel like going to the mountain and falling down into the river and washing ourselves away.

F: Do you know anyone who has killed herself?

Amuloke: There was a woman who had a daughter. Her husband was always angry. So she went to the river and threw herself away with her daughter. Both died.

F: And the husband?

Amuloke: The wife's family asked the man to pay in pigs. At the time the man had five pigs only. The man paid five pigs for the daughter and should pay ten pigs more for his wife.

F: I want to know why your fingers have been cut off.

Amuloke: My older sister died and my mother cut them off when I was five years old with a sharp stone axe—all of them at once. Now I feel a bit angry with my mother because she cut them. When I see the other fingers complete, I feel bad about it. The cut fingers aren't good for holding. They don't work very well.

F: Are your daughter's fingers cut?

Amuloke: My daughter has all ten fingers. I told my daughter I wouldn't do this to her because it causes such suffering in my life for working, for holding. Now younger women and men say it's much better that we stop this.

F: So it's not done anymore?

Amuloke: It's not usually done. The government and the churches don't allow us to cut the fingers.

As Amuloke talks to me, she weaves a string carrying bag—her damaged fingers deftly putting vegetable fibers through their paces. She is amused when she hears about Western distaste for insects. "They have a good flavor," she says, "especially the cicada. It's better than pig."

It's hard for me to make the cultural leap from bug killer to insect gourmand. To regard these most loathsome of creepy creatures as food is a giant step for me, culinarily speaking. I swat flies, step on ants, and cower in the presence of spiders—and never once have I ever thought to eat any of them.

Peter made the leap long ago—he'll eat anything. I, however, must mentally prepare myself before venturing forth into new food territory.

Part of me wants to refuse the insects here in Irian Jaya as I did the fried beetles that Peter ate with gusto elsewhere. But the greater part of me wants to be as adventurous as he is and therein lies the tale: I am doing something that I'm not programmed to do—I'm eating bugs.

If one must, it's advisable to begin by eating insects that crisp up well when roasted. I wouldn't suggest starting with anything too chewy, like a worm, or too fleshy, like cicadas. You want to ease into the experience while not making a total fool of yourself. It's helpful if the people with whom you are feasting are under the age of ten. They will be paying more attention to the meal at hand than to you.

Stink bugs fit the category of crispy insect. Leafy packages, filled by the Soroba village children with the flat hard-shelled bugs, are tossed into a small fire. When the leaves blacken, the bugs are crispy and hot.

Peter takes one and eats it immediately. I take one and find all eyes on me. The children watch intently as they munch their high-protein snack. I am the dinner show—so much for my theory on dining partners. I select a bug and bite down. My mouth fills with a rather bitter medicinal taste, but the charred parts make it palatable. The taste experi-ence is rather like eating a bitter sunflower seed, shell and all, without salt. I chew quickly. It's not terrible and I'm feeling rather proud of myself—for a moment, any-way. My intrepid husband is, of course, pressing on. He has removed one of the roasted spiders from the embers, broken off its charred legs as he saw the children do, and is about to eat it. I studiously ignore the other spiders on the fire.

China

Walking through a market in the southern province of Yunnan, a preadolescent Buddhist monk proudly displays his fake pager—a coveted toy from "Motorora." *Inset:* Water beetles marinated in ginger and soy sauce with carrot garnish against a background of swimming water beetles.

China

July

FAITH: The People's Republic of China, a country of more than a billion people, is a fascinating study in regional contrasts but one facet of life here is remarkably similar in each of its cities: the deafening noise level. The overflow crowd of diners at the Nan Hei (South Sea City Seafood) restaurant in Guangzhou (**opposite**) is oblivious to noise in the dining room rivaling that of an NCAA basketball final.

The similarity to anything Western ends there. This restaurant, like most in southern China, caters to the Cantonese penchant for seeking out the absolute freshest food possible. It proudly displays the catch of the day, live, in two long rows of white plastic crates in the dining room. The setup looks something like a fish market, if the market you imagine also sold flesh-colored marine worms (**preceding pages**), plump pink silkworm pupae, and shiny black hardshelled water beetles (as food—not bait). It comes as no surprise to me that the Cantonese are renowned, even throughout their own country, as adventurous eaters.

We're accompanied by Professor Lu A Ping, an entomologist at Zhongshan University, who has kindly spent the day schooling us on the ins and outs of Chinese bugs. She explains the menu choices and makes suggestions as we order several dishes, including river-worm-and-egg casserole, silkworm pupae stir-fried with fresh veggies, and fried scorpions served on a bed of crisp rice noodles.

I avoid the agony of taking a bite of anything by hiding behind the video camera— "doing my job," I say, as Peter and the professor invite me to join them in sampling the dishes. By the time I work up the courage to put anything in my mouth, the food is cold. I grab a silkworm pupa with my chopsticks along with a piece of green pepper. The pupa pops in my mouth rather unpleasantly and has the consistency of rubber but the taste isn't too bad. As I wash it down with a mouthful of green tea I realize that it might have tasted all right if I'd eaten it hot.

The worms in the egg casserole have no taste of their own and have the same consistency as the eggs. They taste bland and they look bland. I don't get the chance to try the scorpions because Peter and the professor have finished them. Darn.

PETER: Before showing us how to pull the legs, wings, and heads off the water beetles on our plates, Prof. Lu spent the afternoon with us in the Qing Ping market, the sprawling, crawling zoo of a market near the river in an old part of Guangzhou. Cats, dogs, deer, turtles, rabbits, and snakes are all for sale— all for eating. There are insects, too, from marine worms to centipedes on sticks and scorpions (both live and dried); boxes of dried bumblebees (crush with salt, mix with hot water, use for sore throat); silkworm pupae pulsating in plastic tubs; and swarms of water beetles, some of which showed up stir-fried on my plate at dinner. Like a crawdad, the beetle is eaten by breaking open the hard exoskeleton with your fingers and sucking out the white insides. They're messy and a lot of work for so little

meat but the soft stuff on the inside of a bug under that hard exoskeleton is what makes them delicious—like a lobster or crab, you work through the shell and are rewarded with a tasty treat.

Our water bug eating lesson comes in the midst of a bug feast that Professor Lu arranges. Faith is videotaping fastidiously so as to have an excuse not to eat any insects. The food is served on more than a dozen steaming plates in our private dining room. Stir-fried with onions, bamboo shoots, and lots of ginger, the three-inch-long marine worms are like chewy strips of portobello mushrooms. On another plate, the worms are tender but unrecognizable in the coriander-topped baked worm-and-egg casserole. Then comes silkworm pupae, both deep-fried and stir-fried with vegetables. When they are hot, the deep-

fried ones are incredibly tasty. Each one pops in my mouth when I bite down, releasing a rush of flavor not unlike what I imagine a deep-fried peanut skin filled with mild, woody foie gras would taste like. Less successful is the stir-fried pupae with vegetables—the corn-syrupy sauce masks the flavor.

The scariest is deep-fried scorpions on crispy rice noodles. Faith keeps asking me what it tasted like. All I can think of is "barbecued bug." But very tasty barbecued bug.

Not everything is insect or bug-oriented. On the table are dishes of boiled shrimp (heads and tails included) and a very rare steamed chicken, also with its head and feet. Faith is not enthusiastic but says that it is a useful tip for diners: with the head on, you can tell what you're getting.

Fungus in Your Tea

The larvae of the ghost moth *Hepialus armoricanus* have a deadly enemy—the parasitic fungus *Cordyceps sinensis*. Beginning just behind the head, the fungus grows into the body of the caterpillar, eventually replacing all the tissue in its skin. It also grows outward, creating an antlerlike extrusion that can be six inches long. The result is used to make a tea credited with medicinal powers. When the Liaoning Province women's track team shattered the world's record for both 10,000 meters (by 10 seconds) and 1,500 meters (by 2 seconds) in 1993, many outsiders suspected that their extraordinary achievements were the product of performance-enhancing drugs. According to the coach, however, the result was due to rigorous training—and caterpillar fungus tea.

FAITH: The huge electric fans shoving hot air through Silk Factory #1 in the city of Suzhou, near Shanghai, are quiet compared to the clattering din of the machines where a team of women are turning silkworm cocoons into thread. I talk to Tao Xiuzeng (**bottom left**) who has spent eighteen years in the factory pulling cocoons from the hot water bath that loosens the silk, stripping off the silk, and tossing the rest of the pupae into a bucket.

Occasionally she brings silkworm pupae home to eat. "I wash them in water and bake them in the oven to dry," she tells us, "then I fry them in oil with ginger, onion, rice wine, and garlic." The women working nearby agree that stir-frying is the best way to cook silkworm pupae (**right**, stir-fried silkworm pupae in Guangzhou like the ones sold live in the market, **top left**).

Tao's immediate neighbor, working across from her in the factory line, tells us that eating silkworm pupae helps cure health problems like rheumatism. "And they taste good," she says, contributing her laughter to the din. Tao's 8-year-old daughter refuses to eat silkworms. "She's afraid of them," her perplexed mother says.

Li Jiangeng, who handles the finished silk thread in the next room, says she doesn't like to eat the silkworm pupae. After further discussion, she admits that she actually has never eaten them. Asked if she would ever eat a scorpion, a popular item in the south, her mouth drops open in amazement. "I'm afraid of scorpions," she says. "If I had to eat something, maybe the silkworm pupae would taste better."

FAITH: We're lucky to find Li Shuiqi, a 26-year-old scorpion seller, and his roommate, You Zhiming, 25, who sell scorpions by day in the Qing Ping market in Guangzhou (**opposite**) and by night tend to 10,000 scorpions in their apartment. They invite us to come for scorpion soup in the evening and we accept. I've got to see what an apartment full of scorpions looks like, even if I have to eat them to find out.

We're met at the door by both Li and an aroma not unlike rotting athletic shoes. This comes not from the live scorpions, which are in two large boxes stacked with clay roofing tiles, but from the dead ones heaped in the corner. Li and You are selling the dead ones to a woman who is starting a business in traditional Chinese medicine after losing her job in a bankrupt government factory. She uses the stench—these are seriously dead scorpions—to

negotiate a lower price with Li as he tastes, then stirs our pot of soup.

After she leaves, with her plastic-wrapped prize held at arm's length, we eat the scorpion soup (**above,** Li on left and You on right, eating scorpion soup). It's succulent but tastes mostly of pork and Chinese dates, for which I am secretly grateful. As usual, Peter is about three bug-filled bites ahead of me, so I grab a scorpion and eat the whole thing in one bite—all except for the tip of the tail—and chase it with beer that is crisp and cold. The scorpion is woody-tasting. I keep it down.

Li tells us that their families in Luoyang, in east-central China, work in the scorpion-wholesaling business, which is big up there. The city houses many domestic scorpion ranches. Of course we can't miss this, so it's off to beautiful Luoyang we go. Our translator, Shen Maomao, has never been there, so we don't know

quite what to expect.

We fly into dusty, hot Zhengzhou City, which is two hours away from Luoyang by car, and though I'm glad we're not working in Zhengzhou, which has all of the charm of a cement factory, it's safe to say that heavily industrialized Luoyang is not a tourist mecca either. Its status as the birthplace of Chinese civilization lends it a certain cachet, but imagine the pollution of Mexico City amidst decaying Stalinist architecture and you have Luoyang. There's lots of construction—it's hard to tell which buildings are going up and which are coming down. The general ambiance can be summed up by taking in the sight of the fence at the local cigarette factory—tall metal posts painted to look like filter cigarettes standing on end.

You's brother, a quiet helpful man who seems out

of place in the seedy backroom atmosphere of the wholesale scorpion business, introduces us to some of the heavyweights in Luoyang's burgeoning scorpion trade. After assuring what seems like half of the province that we're not trying to steal their business, we're introduced to Hou Songfeng, an eager player in China's emerging market economy. Hou was a traditional medicine doctor, using the scorpions he raised in his home for treatments. He has parlayed his knowledge of scorpions into a joint business venture with a community outside of Luoyang. The new business is the Ru Yang Boda Scorpion Breeding Company. With thirty employees, he's raising three million scorpions in a brick facility the size of a football field (**following pages**).

Scorpions are fairly quiet, but when we go in, we can hear the scuttling of millions of tiny feet. I'm not exactly sure where they are until my eyes adjust to the darkness. (My nose never gets adjusted.) It's unnerving to be in this giant, almost lightless building with three million poisonous creatures. This place is made for scorpions, not people, and scorpions like darkness and dankness and other scorpions.

Reassured after a bout of companionable tea drinking that we aren't trying to steal his trade secrets, Hou serves as an enthusiastic guide through his operation. He wants to open up new markets and wants to know if anyone in the United States of America is interested. Hoping that our visit will somehow translate into overseas investment, he gives us his phone number.

FAITH: We've heard that many insects are eaten in Kunming—the capital of the southern province of Yunnan—so of course we must travel there, even though it means taking another China Airlines white-knuckle special. We make our way to the city and within hours find ourselves in a tenement-style walk-up, talking with an antique dealer about his collection of period Mao posters as we wait for his wife, Fan Yuelian.

I had liked the look of her as she cheerfully arranged bright-colored seed necklaces on a display above her buckets of beetles and scorpions in the city's crowded Bird and Flower Market. While we talked I watched—fascinated—as she absently turned her hand through the bugs, allowing the hard-shelled beetles to crawl over her fingers and fall with a clatter onto their fellows in

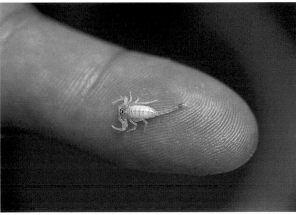

her plastic tub. Fan Yuelian said she'd been raising insects with her husband's family for about ten years, mostly for medicine but also for food, and invited us to come to her home.

Later, in her apartment, we follow her into her son's bedroom, which is draped prettily with mosquito net and bright-colored Shanghai sheets. Stacked by the bed are plastic bins of live scorpions (**top**), masses of them. There are all sizes, from

baby to elders (**above and right**).

"Your son doesn't mind sharing a bedroom with a scorpion ranch?" I ask, imagining our sons' reaction to this arrangement.

"Why would he mind?" Fan asks. "They can't get out." Why indeed, I think. Though the room's gamy aroma would be a point against them, this is a family business and in China that means everyone has a role.

PETER: In Luoyang, we order scorpion soup and stir-fried scorpions at the Elegant Good Smell restaurant, where the AC is off and the smells are neither elegant nor good. While they are being prepared (elegantly, we beseech, since we will photograph them before we eat them), I go into the kitchen. Against the wall, inside a stainless-steel hood, are brick blast furnaces directed through stainless steel holes onto huge woks (**above**).

When I get back, the chef and manager appear with a special treat—an appetizer of live scorpions. First the manager puts them in a small bowl of water. The scorpions aren't happy about this—they start thrashing about. A good sign, I decide. If we are going to eat live scorpions, let them be very alive. With chopsticks, the manager removes the scorpions from their bath and

drops them in rice wine for a few minutes. The scorpions stop struggling and go into a coma. The chef then scissors off the tail stingers and poison sacs and arranges the scorpions on a plate.

After I photograph the chef eating a few, it becomes clear that it's my turn. We've been in China for a month and so far my taste buds have been assaulted more times than the Great Wall. I brace myself, but the experience isn't so bad. It's very chewy with a gutsy, almost fishy taste, but it's overpowered by the wine flavor. They're much better fried, so I only eat a couple.

That evening we go to the Yue Xiu seafood restaurant. When we walk in, we are amazed. The place is elegant and upscale, surprisingly unlike the rest of Luoyang. It reminds me a little of a European bistro, except for the jars of herb, deer penis, and snake brandies on the bar. Wang Lingyun, the manager (**above right**), is a

refined young woman who seems completely out of place in this industrial backwater. Scorpions are a fixture on the menu—the restaurant is famous for them. People like them for snacks, she says.

We try the scorpions she serves (**below right**) and there's something to be said for ambiance and presentation. The scorpions arrive with a fancy garnish of maraschino cherries and radishes sculpted to look like butterflies. The scorpions themselves taste just like the others—like sautéed twigs, but good ones.

Good News, Bad News

The reputation of scorpions as hazards to human life may be overblown; of the many species in six families worldwide, there are really only a few that can be considered dangerous. But if you do get stung: In mild cases of scorpion sting, effects can include swelling, pain, and discoloration at the site of the sting; fever; and occasionally nausea and vomiting. Such symptoms usually subside within about twenty-four hours. In more severe cases, symptoms include anxiety and agitation; severe pain at the site of the sting; excessive salivation and perspiration; an irregular pulse and an unstable body temperature; breathing difficulties; muscular twitching, which in its extreme form leads to convulsions and death. Examination of fatalities often reveals damage to both the heart and lungs.

Stinger Zinger

Overuse of pesticides has reduced the wild population of *Mesobuthus martensii*, the only scorpion common to China. The decline is a small boon for public health, because scorpion stings are dangerous. Although the venom rarely kills adults, small children and especially infants are vulnerable. When scorpions are eaten raw, the cook must remove the stingers and venom sacs because the stingers are extremely sharp and the toxins could be absorbed through the mucosal membranes (an initial sign of trouble is a Novocaine-like taste and sensation on the tongue). Cooked scorpions are not devenomed, probably because the small proteins that comprise the toxin are broken down by the heat of cooking.

FAITH: "Winter worm, summer grass" (*dong chong xia cao*) is what the Chinese call the caterpillar fungus sold in markets and traditional-medicine dispensaries throughout China. The fungus literally takes over the insect's body; the caterpillar becomes a sort of fungus sculpture of a caterpillar. It's hard and brown, a bit mummylike. The result is used to treat asthma, colds, jaundice, and tuberculosis. Laboratory research has found that fungus-infected caterpillars have several pharmacologically active components, including some with antitumor properties. Like ginseng, the fungus is regarded as what I will delicately call a "strengthening agent."

Caterpillar fungus is *expensive*. At the Qing Ping market in Guangzhou (**above left**) the price for one pound is 4,000 yuan ($500 U.S.). We try it at the Wine Forest restaurant in

Kunming, a place highly recommended by our guide books for its healthy dishes. A small pot of Caterpillar Fungus Soup with Black Chicken (**above right**) costs 50 yuan ($6.25 U.S.), exorbitant by local standards. This is the price despite the fact that there are only three or four fungi in the soup.

Someone worked very hard on the aesthetics of the Wine Forest. Menu items, for instance, are written down on individual pieces of bam-

boo stuck in a pot. Patrons flip through them to decide what to order. Too bad the attention to detail didn't extend to attitude implants for the waitresses who work there. The manager is pleasant on the phone and invites us to come but is gone when we arrive. No amount of talking will budge the forbidding staff who refuse to allow Peter to photograph the restaurant even though

we are obviously not disturbing any diners. There aren't any.

The staff is incensed when Peter photographs our meal of caterpillar fungus soup, and their blood boils when they see my notebook. Our waitress glares at us and tells the others that we're stealing state secrets—apparently forgetting that we are accompanied by our translator, Maomao, who not only hears this ranting diatribe but translates it with glee. I'm not sure what the state secrets could be, but we walk out knowing that the other ingredients of this soup are chicken broth, dried spices, onion, garlic, red pepper, and chili pepper.

PETER: Huge hotel and restaurant complexes are under construction everywhere in Jinhong, near Burma, creating a boomtown feel. Chinese tourists are thronging here. So are prostitutes. After a long day spent in nearby villages like Menghan, whose Sunday market (**following pages**) draws hundreds of colorfully dressed indigenous Dai and Atsch peoples from the area, buying and selling everything from tropical fruit to water buffalos, Maomao and I walk to the night food market on a muddy street leading to the Mekong River. There, a dozen stalls offer pig brains, pig feet, chicken feet, dried frogs, fish heads, and—eureka!—dried cicadas (**bottom left**).

While we watch, Yue Ta,

the cicada saleswoman, de-wings a dozen and cooks them in a wok for 40 seconds. They're surprisingly good, even though they are dried, and have a crispy mild taste like salty bacon puffs. I convince Maomao to eat two. I eat the other ten, washing them down with an icy cold Mekong beer. Back at the hotel I have to rinse my mouth before brushing my teeth. The sink is littered with cicada body parts that were stuck between my teeth. It looks like a scene from a dentist's nightmare.

Feeling Frassy

Chongcha is a tea made from the frass—the technical term for insect excrement—of *Hydrillodes morosa* and *Aglossa dimidita*, two species without common names in the West. The former is a noctuid caterpillar, a member of the family Noctuidae, which includes the cabbage looper and corn earworm. *A. dimidita* is a member of the family *Pyralide*, the snout moths. In the mountainous areas of Funan, Guixhou, and Guangxi, the excrement is highly prized for *chongcha*. This black fragrant beverage is thought to aid digestion, alleviate diarrhea, and treat bleeding hemorrhoids. *Chongcha* has been little studied by Western science, but its medicinal properties may well be due to the presence in the frass of pharmacologically active substances in the insect's diet. Such compounds often pass through the caterpillar's digestive system unchanged and are excreted in the frass. A well-known example is the South American malumbia caterpillar, which feeds exclusively on *Erythroxylon coca*; its frass contains substantial amounts of pure cocaine.

PETER: "Good Smell Bad Taste Road" could be the name of the new four-lane highway in Kunming that's lined with monstrous theme restaurants—cyborg versions of Planet Hollywood and the Hard Rock Cafe. Looking like a warehouse gagging on a fake ship, one huge establishment has the stern of a Spanish galleon emerging from the front. Caged ostriches molt outside another restaurant—a faux castle complete with moat. For lunch we choose the Roasted Goose restaurant. Two huge plastic geese flank the palatial entrance. Inside the cavernous interior, more than 200 waiters and waitresses in ill-fitting green uniforms with paper goose hats stand at semi-attention.

Separating the kitchen from the dining area is a grocery-style refrigerated display case with every conceivable edible Chinese delicacy. We go for the ant and chicken egg casserole (**above**). The formic-acid ant flavor permeates each bite with a sour taste that is not unpleasant. But the crunchy black ant bodies clash with the fluffy eggs: It's like eating pencil shavings in a pudding.

Along with the ant casserole, you can drink ant wine—ant-steeped rice brandy, actually. You make it by steeping ants in rice brandy. The formic acid and minerals in ants are said to be effective against hepatitis B and rheumatism. The wine is so popular that the medicinal ant genus, *Polyrhachis*, is in danger of extinction. According to *The Food Insects Newsletter*, commercial anthills—formicaries—are being developed.

Hoping to get the farm-raised variety, we buy a kilo of black ants and a bottle of ant wine in Shanghai's Insect Research Institute. We take them to Beijing, where we hire a car to take us to the Great Wall at Mutianyu. We walk along the wall in the

July heat and after two hours settle on a mountainous section of the wall where Faith arranges the ants (which were blowing around) and the wine (**far left**). Afterward, we sip the ant wine, which seems like the sensible thing to do since our knees are aching from the climb.

FAITH: The traditional-medicine dispensary, Fu Lin Tang Pharmacy, in Kunming is near the Bird and Flower Market. Customers stand in a long line that snakes around the room waiting to discuss their ailments with the traditional-medicine doctor (**above**). Those near the front of the line get to hear the questions of the patients in front of them before it's their turn to get their maladies overheard by the patients behind them. As always in China, no one appears to mind the total lack of privacy.

After an analysis of the problem, a treatment is prescribed and the order for medicine carried to the back of the dispensary—a floor to-ceiling treasure trove of old wooden drawers polished smooth by years of use. Women in white coats fill the doctor's orders using ancient scales. The room smells comfortably of ginger root and anise seed and other scents I can't identify.

One remedy, being studied at a local university in Kunming by Professor Chen Xiaoming, is caterpillar castings tea—called *chongcha*. (see "Feeling Frassy" on page 101).The tiny hard pellets of caterpillar excrement are dumped into hot water and steeped (**left**). The result is an excellent example of the dictum that people don't think substances are medicinal unless they taste bad. I didn't want to drink this tea but did so in order to tell you that it tastes like dirt—which makes sense, if you think about it.

Mexico

Sixty miles southwest of Mexico City, schoolchildren in the town of Taxco celebrate Jumil Day, a festival in which crowds gather for the ritual harvesting and eating of *jumiles* (a type of stink bug). *Inset:* Roasted grasshoppers and mashed avocado on a corn tortilla.

Mexico

August

PETER: When Faith and I returned to Mexico, we went to visit Julieta Ramos-Elorduy, the culinary entomologist in Mexico City, and the memory of the first live stink bug that died between my teeth came rushing back. I'd gone to the crowded Jumil Day festival on a mountain top southwest of the capital, where Julieta explained that in Mexico insects are regarded not as the last resort of poor people warding off starvation, but rather as delicacies to be served to honored guests. Having invited myself to the festival, I was not exactly an honored guest, but I intended to eat a *jumil* anyway, *Euchistus taxcoensis*. Julieta cautioned me that stink bugs might shock my taste buds. She neglected to warn me, however, that this nasty-tasting little creature would fight back. Humans ended on top

of the food chain not only because of our size and intelligence, but because of our quick reflexes. In other words, when the jumil started clawing at my tongue I reacted with a reflexive *chomp!* This chomp was how I discovered the second thing Julieta had neglected to warn me about: Stink bugs have a very strong taste—like an aspirin saturated in cod liver oil with dangerous subcurrents of rubbing alcohol and iodine (**photo, page 15**). I can see why they are usually ground to a paste with chiles and tomatoes—*chiles de jumil.* Despite Julieta's words I couldn't see this as the main course for honored guests— more like the kind of crazy food that you might gulp down once a year at a wild party, like goldfish or chocolate-covered ants.

If stink bugs are strictly a once-a-year treat, the grasshoppers we found on this second visit to Mexico are daily bread. Because of their numbers, wide geo-

graphic distribution, and ease of collection and preservation, *chapulines,* or grasshoppers, are eaten all over the country (**above,** an unusually large example). They are especially popular in the southern state of Oaxaca, which is where we went to look for them.

FAITH: Leisurely Oaxaca is a hugely different experience than polluted, crime-ridden, cosmopolitan Mexico City.

Zapotec and Mixtec Indian people still predominate in this region. Their ancient ruins are a big tourist draw, and their art is now an important business, but for most people, life in the villages continues to be fairly simple. The countryside is rich in traditional adobe houses, cactus fencing, and colorful flower gardens.

It's here that we find Irene Martínez Pablo and her sisters selling *chapulines* in the market for ten pesos ($1.25 U.S.) per large cup. She tells us she catches grasshoppers near her village of Santa Lucía Ocotlán, just outside town, and that we can join her if we like.

At daybreak the next morning, we walk with Irene and her nieces and nephews to a good grasshopper spot "just over there." This becomes an hour's tromp past newly harvested cornfields (**preceding pages**). As we follow with cameras ready, passersby tease Irene about becoming a movie star. Grinning widely, Irene flips her braids about, tying them together in Zapotec fashion, with cotton strips of cloth.

We walk off the path to a field of tall grasses browned by lack of rain. The kids chase the grasshoppers with their hands while Irene drags a bag along the ground, trapping the insects as they jump up. Once home, she pan-roasts the grasshoppers. Like lobsters, they turn red when cooked. Irene tosses them in lemon, salt, and garlic both for flavor and because this preserves them for the market. Back in our room in Oaxaca, I set up a photograph (**right**) using grasshoppers and a small clay figure sculpted locally. After photographing this food display, Peter eats it.

FAITH: A half-hour drive outside Mexico City within sight of the ancient Pyramid of the Sun at Teotihuacán, we screech to a dusty halt. The roadside cactusfruit venders barely look up. Julieta, our entomologist guide, has taken us here to buy red agave worms. We're in luck; several men and women are offering plastic bags of worms for sale to every car that stops. The price is ten pesos for a small bag of ten inch-long worms. But the vendors don't want to talk with us. Is it because we're foreigners? No, they suspect Julieta is trying to steal their business. Julieta calms their fears, explaining that she is a bug researcher, but this leads to a different

problem as the vendors clamor for her help in creating a bigger worm business.

Seller (who refuses to give his name): Many people now want to eat this red agave worm but they don't want to pay the amount we ask. This product is sold at very luxurious restaurants. The value of this insect is not by the monetary price. It's by the flavor price. We are not asking too much. We buy the worms for 280 pesos per kilo and sell on the road for 320 pesos so there is only a gain of 40 pesos (about $5.00 US).

Not everyone eats them, the vendor says. People are sometimes "afraid" when they see worms for sale as food; or ants, which he sells in the springtime; or grasshoppers in the rainy season; or white agave worms.

Faith: Why do you think that some people are afraid of insects and others eat them?

Seller: If I go to a restaurant around here on Christmas Day and ask for oysters, they might be astonished and disgusted that I would ask for something like that. But if I go to Veracruz on the coast, all the people there know about oysters. The same thing happens with the worms. If the people don't have this habit of eating something, they will feel repulsion about it and will not buy it.

F: How do you like to eat the worms?

Seller: I eat them alive, roasted, fried, or with other foods. It doesn't matter. I also like to eat them with chile sauce.

Platters of cooked grasshoppers are sold everywhere in Oaxaca. Paco, the grandson of the woman whose guest house we stayed in, had *chapulines* after school every day with his grandmother (**top left**). Women carry heaping trays of grasshoppers around the city marketplaces and some also sell maguey worms and a seasoning made of salt, chiles, and crushed worms. The maguey worms are better than the grasshoppers, which taste a bit rancid. "Is there a difference in taste between the different sizes of grasshoppers?" I ask Rosa Matíaz, who sells *chapulines* and maguey worms in Oaxaca's Central Market (**above**). "The smallest ones are the best," she says.

Insects are sold in the Abastos market as well (**lower left**) though they're only one of the attractions. We pass three young cricket vendors totally engrossed in their books, which have half-naked people on the covers. I ask them, Are those trashy *novelas* you're reading? There's laughter and quick hiding of the paperbacks, but no reply.

Bug Proof

Why do distillers put caterpillars into tequila bottles? Typically, they use maguey worms, the larvae of giant skipper moths (*Aegiale hesperiaris*), which feed on agave plants. The caterpillars certify authenticity, because real tequila and mezcal are produced only from agave, and the only place these caterpillars exist is in agave plants. Equally important, they attest to the potency of the brew. Tequila and mescal are traditionally double-distilled to at least 110 proof, but are sometimes unscrupulously watered down. Caterpillar bodies cannot be preserved for long at less than 140 proof. Perfectly pickled caterpillars in bottles are thus evidence that the contents have not been adulterated.

PETER: Today Juan Cruz and Pedro Mendoza find only a few small red agave worms while cultivating their maguey cacti near Matatlán, an hour south of Oaxaca on the high, dry plains of mezcal and tequila country (**preceding pages**). These cactus plants are young, just three years old. In another few years the heart of each cactus will weigh nearly one hundred pounds—enough, Juan says, to produce five quarts of mezcal, the smoky brother of tequila. If it had rained the night before, Juan tells us, the worms would be crawling all over the ground— easy picking, and good money, too, because he can sell the worms to the local mezcal *cooperativa* (**right**). The worms that don't end up in mezcal bottles can end up on dinner plates (**far right**, *Chinicuiles con Aguacate:* fried red agave worms and small corn tortillas, with refried beans, grated cheese, sour cream,

and avocado). At the Restaurante Zempoala, near Teotihuacán, both red and white agave worms are on the menu, along with *escamole* (ant larvae and pupae). We eat all these insects, but the best are the fried red agave worms. They are as fresh as the ones squirming in the roadside seller's bucket (**top left**).

FAITH: Over lunch, one point becomes crystal clear: María Luisa Aguirre del Gadillo (**above**, frying red agave worms), who owns the Restaurante Zempoala, wants us to sell her insects in the United States. Our friend Julieta dismisses the idea—import rules forbid it and besides, the bugs will spoil. This is the end of the discussion as far as we're concerned, but it's just the beginning for the señora, a budding entrepreneur with a freezer full of ants and agave worms to unload. She has acquired them from friends and family to make a killing

in the marketplace, just as soon as she learns where the marketplace is. Throughout the afternoon she regales us with the commercial possibilities. What kind of commission are we interested in?

Opening the freezer in the back of the restaurant, she shows us her stash: 770 pounds of ants and 440 pounds of worms, maybe $15,000 worth of frozen larvae. An amazing commercial miscalculation: She can't possibly use them all. I've never seen so many frozen bugs at one time—actually, I've never seen any. The señora, resilient in the face of potential loss, keeps probing. Having concluded that we can't or won't help her directly, she hits on another idea. "Do you know anyone from other countries who might want to buy them?" she asks. This woman's tenaciousness would serve her well in China dealing with the scorpion wholesalers.

PETER: A short time after Julieta Ramos-Elorduy had me eat stink bugs, she spent a weekend cooking up dishes like mealworm spaghetti (**left**) and stink bug paté (**upper right**) in her Mexico City kitchen. A researcher who brings her work home, Julieta has a refrigerator that is a science project in itself—dozens of containers of live and dead insects. The insects are part of a cookbook project with dozens of bug recipes that she has collected or concocted herself. She is cooking them with the help of a French-Polynesian chef and a food stylist from San Francisco, and she asked me to photograph the results. She works all weekend and I shoot dish after dish: garlic crickets; mango-grasshopper chutney; white agave worms in white wine; Mexican caviar (eggs of lake-water boatmen and backswimmers); blackwitch moth larvae salad; fried tree hoppers (harvested from an avocado tree next to her university office); fruit salad with wasp honey; mescapale (aquatic larvae of predatory diving beetles) tamales; escamoles (ant larvae and pupae) with chiles; brochettes of longhorn beetle larvae; and flan Chicatan (egg custard with leaf cutter ants). After photographing each dish, we eat it. There are too many for us to finish them all but we eat enough to get a good taste. The dishes are delicious—Julieta is a very good cook. Nevertheless, as she admits, she has had little success in her goal of reviving a pre-Columbian insect-focused diet. "People want to eat what they see on TV," she laments. She has made one convert, though. Late on Sunday night we are finishing up when Julieta's youngest son, Ernesto,

comes downstairs from his room, hungry from an evening of college homework. Since we have a big container of leftover leaf-footed bugs, Julieta agrees to make Ernesto his favorite dish: leaf-footed bug pizza. An hour later what comes out of the oven (**lower right**) is straight out of an Italian chef's nightmare. But the pizza is delicious.

Stink Bug Paté

½ lb. roasted stink bugs (*jumiles*)
10 chicken livers
4 cloves garlic
1 small onion
⅛ teaspoon salt
Ground black pepper
Ground oregano
Ground marjoram
Powdered bouillon
Olive oil, to taste

Boil chicken livers, garlic, onion, and salt in enough water to cover. Simmer 10 minutes. Run livers through food processor, reserving the broth. Add roasted bugs to livers; blend. Add the broth until the mixture has the consistency of a thick sauce. Add spices and oil to taste. Form paté, serve with French bread.

Mealworm Spaghetti

½ lb. roasted yellow meal worms
4¼ cups water
1 tablespoon safflower oil
1 sprig marjoram
1 sprig thyme
2 bay leaves
¼ onion, chopped
8 oz. dry spaghetti
6-8 tablespoons butter
¼ teaspoon salt
Olive oil
3-4 tablespoons pine nuts, finely chopped.
10 sprigs parsley, finely chopped.
½ lb. purple basil, finely chopped.
½ lb. ricotta cheese
¼ cup whole pine nuts

Boil water; add safflower oil, salt, marjoram, thyme, bay leaves, and onion. Add spaghetti. Drain when done. Melt butter in sauté pan. Add spaghetti. Salt and pepper to taste. Mix basil, parsley, ricotta, oil, and chopped pine nuts with the spaghetti. Heat but do not boil. Top with mealworms and whole pine nuts.

PETER: For sixteen years, Cresenciana Rodríguez Nieves, a 43-year-old doctor in the city of Puebla, at the base of the snowcapped Iztaccíhuatl volcano in central Mexico, has used a wide variety of native plants, animals, and insects in her practice of what she calls "México" medicine (**above**). "This," she says, "is the medicine practiced here before the Europeans came." (She doesn't like the term "traditional medicine," because she regards it as a pejorative term used to dismiss remedies based on centuries of folk experience.) Her expertise as a medical doctor has helped her to appreciate the wisdom of, say, treating goiter with stink bugs (they are rich in iodine, and goiter is caused by iodine deficiency). She treats anemia with grasshoppers, rheumatism with beeswax, eye cysts with flies—the range of medical applications is enormous,

she says. After an impromptu lecture, Cresenciana graciously serves us a delicious lunch of carrot soup and fresh corn tortillas (no bugs) and then shows us how to catch grasshoppers with a big woven basket in her backyard.

FAITH: In Oaxaca, we stop chasing edible bugs long enough to explore the works of skilled weavers in the area—and we find more bugs. Female plant-feeding cochineal scale insects (*Dactylopius coccus*) are dried and crushed by the weavers and used in dye baths to give their yarns a vibrant red color. The cochineal is also used commercially to color commonly used items like dental plaque remover and sugarless candy, and more exotic items like the Italian liqueur Campari.

Lest anyone wonder for too long why only the female cochineal is used—wonder no longer. The female cochineals live in small

groups and eat prickly pear cactus. Males are solitary and eat nothing, because they have no mouths. They only live for about a week, during which time they grow wings, fly off to mate, and die.

I like to think of this as the work of Nature in her imponderable wisdom.

To make half a kilo of dye, 75,000 insects are harvested by hand and plunged into boiling water to remove their protective coating. Then the bodies are ground into a powder. The result is a color that was so widely treasured in the late fifteenth century that the Aztecs demanded it as tribute from conquered tribes in cochineal areas. (In the Middle East, red dye from another scale insect, *Kermococcus vermilius*, was almost as valuable; it was probably used by the ancient Sumerians.) After the Spanish conquest, cochineal became a profitable Spanish monopoly. When Spain

banned the exportation of live cochineals, French and Dutch bug bootleggers tried to smuggle them out. They were unsuccessful until the nineteenth century, when France created a cochineal industry in then-colonial Algeria. Peru and the Canary Islands also became big producers.

Cochineal dye was replaced on the marketplace by synthetics until the recent resurgence of interest in natural products. Today many Oaxacans (**right,** the weaver Benito with a mortar of ground cochineal) have come back to cochineal to give their weavings the prized traditional look. Not everyone is happy going back to natural products, though: Lipsticks made with dyes from insects are not considered kosher by Orthodox Jews.

Botswana & South Africa

Late afternoon laundry shadows play on the courtyard of a typical family compound in the village of Masetoni in Venda, in the far northeast of South Africa. *Inset:* A simmering pot of mopane worms in Botswana.

Botswana

January

PETER: In a race against the clock, we drive like mad through the former South African "homeland" of Venda because the border with Botswana closes at 4 p.m. We can only get a dusty car-window-view of this northeastern corner of the country as we bullet past landscapes that vary from scraggly brush to lush river valleys to mountain forests.

The border at Zanzibar becomes more like backyards than backwoods as the gravel road narrows to a dirt track. The crossing itself, over the swollen Limpopo River, is a low one-lane bridge with a one-room concrete shed for the border guards. They're friendly and we arrive in time, but we're lucky to have Afrikaner Ben van der Waal, the son of Dutch immigrants, guiding us through the formalities. Ben is a biologist at the University of Venda in the Northern Province (former Northern Transvaal). After exchanging e-mail for several months, we showed up at his house after a five-hour drive from Johannesburg and became instant friends. At the drop of a hat he loaned us camping equipment, which we crammed into our rental car before speeding north to Botswana.

After the border, we stop at every roadside shop, and Ben asks about mopane worms, the edible caterpillars that are a favorite in southern Africa (**preceding pages**). The people we're looking for are harvesting worms to sell in city markets or to wholesalers, or to keep for themselves to eat. Dried mopane worms have three times the protein content of beef by unit weight, and can be stored for many months.

Grace Mapitse's Sun-Dried Mopane Worms

3 cups fresh worms
1 cup salt

Align 4 or 5 worms in both hands. Squeeze like milking a cow, discard liquid guts. Put worms into pan with salt and water to cover. Boil on fire until water evaporates, about 30 minutes. Spread worms on empty bags in sun, turning every hour until dry, about 1 day.

By sunset we find mopane pickers camped off the road eighty miles north of the border between Selebi-Phikwe and Francistown. Yes, they will pick worms tomorrow; we promise to return at 6 a.m. With ice, beer, a frozen chicken, a tarp, and some sleeping mats, we are ready for the evening. We pull onto a gated ranch road, drive a mile into the mopane scrub land, and make a fire. In one kettle, Ben cooks a great dinner of curry chicken; in another, he whips up some porridge. We're asleep under the stars by 10 p.m.

FAITH: On the grassy, open veld nothing bothers us but the stars shining on our faces. After the mad dash to the border, there's finally time to reflect on the mopane worm harvest that we've been chasing. We first scouted in South Africa, but the sellers in Venda (**above**) said there hadn't been enough rain for a good crop of worms. Their dried worms come from southern Botswana, so here we are.

Harvest time is a campout. Whole families leave their homes and converge on government, private, and tribal lands, where they build temporary housing from plastic sheeting, canvas, and even roofing paper. Each family's compound looks like a small, bustling tent city. The tools of the harvest are simple: rubber gloves, buckets, cooking kettles, salt for preserving, and sheets of canvas and plastic bags for sun-drying the cooked worms (**left**).

In the camp we meet Grace Mapitse and her group of about twenty women and children. They've been here for a month in an area full of mopane trees. Twice a day, Grace tells me, they gather, clean, salt, simmer, and sun-dry heaps of mopane worms. As evening falls and this hot January day grows cooler (we're below the equator so it's summertime here), Grace adds the day's batch of worms to the family's growing stock, fishes money out of her brassiere for a cousin, and yells at another cousin to check the progress of mopane drying at the edge of their camp.

FAITH: As arranged last night, we meet Julia Marumo, her two young sisters, and her cousin Gladys as they leave their camp to pick caterpillars. The air is fairly cool but promises to get hotter. As usual, it's difficult to get started. The girls are cramped by our closeness, and moving pictures and stills interfere with each other, so I hang back a bit with the video camera and my micro-recorder and watch the two youngest girls stare into Peter's camera. They don't speak English, one of the official languages here, and I can't speak their language, Setswapong, so I can't tell them to ignore him. Julia senses what we need and speaks to them in clear, quiet English. Gladys's English isn't as good but she loves our attention and so seems more friendly.

It's Julia I'm drawn to though. In her subdued and focused manner I can see a sadness that comes from somewhere I haven't yet discovered. "We haven't been here very long," she says. The harvest isn't going as well as they had hoped, so they will probably only stay a few days more. Julia has arranged for a truck driver to take her family home to her grandmother's house in Leruli, not far away.

I half expect Julia's two young sisters to make a game of picking the worms

and when bored, to run off, but they don't. These caterpillars mean business.

The mopane worm is actually the caterpillar of the anomalous emperor moth (*Imbrasia belina*), one of the larger moths in the world. "Mopane" refers to the mopane tree, which the caterpillar eats. Worms are harvested twice a year and they're a very big deal. The caterpillar, a cash crop, is in great demand, especially in South Africa. In fact, over-

exploitation is threatening the caterpillar with extinction, so South African entomologists are trying to develop methods to farm-raise them. Harvesting mopane worms provides income for many people, but it can't keep going like this unless someone figures out how to sustain the caterpillar population. Catching mopane worms is messy and hard on the

hands. The worms have thornlike points on their backs that are sharp enough to slice unwary fingers. Before beginning work, Gladys pulls on a pair of ribbed rubber gloves, standard worm-harvesting gear. All that remains of Julia's gloves is a handful of rubber fingertips. She has to ask her sisters to help her pull on each fingertip.

At the mopane trees the women and girls work together picking off all of the larger worms (**left**). Julia grabs a handful, and holding them tightly at one end, she squeezes out the insides from the other end. Bright green and yellow juice spurts out—instant death for the caterpillar—and then the worms are tossed into the bucket. The guts smell like freshly crushed leaves, which is exactly what they are. The cycle—pick, squeeze, toss—happens over and over, filling the buckets to capacity as the day heats up. At first the spray of yellow-green slime nauseates me; then I get used to it. Inevitably the pickers' clothes get stained with the bright caterpillar juice that, when dry, turns brown. No wonder everyone's wearing rags.

"You're not taking the smaller ones," I comment as I watch Gladys work her way around a mopane tree. "They are not as good. The fat ones are better," she tells me. Ben van der Waal, our biologist friend, explains, "The younger worms don't have as much fat and they are not nice to eat. The fat is actually what gives them the nice taste and this develops only just before they pupate." In fact, he says, the comparatively unpleasant taste of young mopane worms is one reason the species isn't already extinct. The bad taste

means that they are left to mature and complete their life cycle. When the sun is directly overhead, everyone takes a short break for a snack of sweet wild plums from a nearby tree. The younger girls clean their hands and feed the fruit into the mouths of the older ones, whose hands are still covered with caterpillar guts.

Back at the mopane trees, I follow Julia, who tells me matter-of-factly that her mother died of a liver ailment in 1992 and her father died of the same thing. I determine later that both parents died of AIDS, and she is now the sole support of her sisters. In addition to this burden, the father of her toddler refuses to support their child. He moved to Johannesburg and she doesn't hear from him.

When Julia is not gathering mopane worms, she helps support her family by brewing homemade beer from sorghum. She learned English on her own, not from school, which she has no time or money to attend. It's her dream to go back, she says, but she doesn't think that will happen. Instead, she works to pay for her younger sisters' education. "They like school," she says, proudly, if not a little wistfully.

PETER: Sitting in the dirt, in the partial shade of a mopane tree—most of the leaves have been eaten by the hungry worms—as Mamebogo, Julia's little sister, squeezes the guts out of live mopane worms (**bottom right**) and carries off the carcasses (**left**), I can't imagine I am watching a major commercial enterprise. But trade in dried mopane worms is bigger than I had thought.

On our way out of Botswana, the Limpopo River is flooding, which forces us to detour thirty miles southeast, to a higher bridge at Platjanbridge. At this border crossing, we meet a young Afrikaans-speaking woman, Alet van der Walt, and her two-year-old son, Walt. Alet's HiLux pickup and trailer are loaded with a ton of gunny sacks full of dried mopane worms, which Walt loves to munch on (**above right**). I think they taste like sawdust. Alet and Walt are returning to their farm in northern South Africa to sell the worms to wholesalers, who resell them to merchants, who resell them again in the market (**following pages**, mopane sellers in the market of the South African city of Thohoyandou) —a lucrative traffic, all in all.

FAITH: Alet also raises heifers, but their selling prices are too low, so she is concentrating on mopane worms. About once a week for the last five years of harvests, she has crossed the border with loads of worms. Sometimes she is able to fill up her truck with the output from a single mopane camp; other times she has to go from place to place, buying a few hundred pounds at a time. I ask her if the trade is profitable.

Alet: Yeah, there's good profit. But it's a tough business, because we must sleep in the bush, and all that sort of thing.

F: How did you get into this business?

Alet: Somebody told me about it, and he said I must try. He was a policeman here and he did it in his spare time, but then he left. So when he said to me I must try it, take over his license, I did.

F: Does your son always travel with you?

Alet: Only when I'm not traveling for a long time. Now that the weather's cooler, I can take him, but when it's hot I can't.

F: How long do you think you'll do the business?

Alet: What I want to do is to go bigger and buy 200 bags or so at once and come back.

She is quite clear about her hopes. Like almost every entrepreneur I've ever encountered, she wants to expand rapidly. There on the South African border, she's thinking about making a play for the big time.

South Africa

January

FAITH: In far northeastern South Africa, thatched circular dwellings called rondavels stand in groups of three to six on the gentle hills. Formerly known as the Northern Transvaal, this area was renamed Mpumalanga and Northern Province after the end of apartheid. Its northeastern region is the home of the Venda people, who have lived in this part of the world since the twelfth century; their houses are round because no evil spirits can lurk in a house that has no corners. Extended families live in the groups of houses, creating communities within communities. It is to one of these family compounds in the village of Tshamulavhu that we've come with Tshifhiwa Munzhedzi, who calls himself "Eric." Eric's stepmother, Muditami Munzhedzi, is going to cook a breakfast of mopane worms for us.

The family hastily produces benches and insists we occupy them, an honor that we try to refuse as everyone else is sitting on the smooth-packed earthen courtyard. There are thirteen family members living here, including Eric's niece, who was abandoned by his oldest brother, and his sister, whom he says is under a spell cast by her estranged husband and has problems in her head. Eric, 29, a new graduate of the University of Venda, is the sole breadwinner, working part-time at his school. His father is retired from working at the nearby Kruger National Park but is not yet old enough to receive a government pension.

In most of southern Africa the staple food is cornmeal porridge, and Venda is no exception. Cooked twice a

day, the porridge is eaten with each meal and as a snack. A freshly cooked plate has been prepared. All that remains is for Eric's stepmother to cook the mopane worms. Muditami has put on her traditional clothes and the slight sardonic gleam in her eye as she looks over at us (**left**) makes me wish even more that we spoke the same language. I'll bet she's a pistol.

Breakfast is mopane worm stew. Muditami reconstitutes the dried worms in a pail of water while readying a small fire in the courtyard. Plumped up by the water, the worms go into a clay pot with a small amount of water and the rest of the ingredients. The children, who have been surrounding us all the while, lick their lips in anticipation.

When the stew is ready, we share it in the shade of a nearby tree (**above,** Eric in white shirt at left). Everyone

takes a dip in the hand-washing bucket then grabs a bit of porridge and stew from the communal bowls. Concentrating on the tomato and the porridge, I eat the worm on my plate. It's chewy, juicy, and salty—but not bad. Once again, I eat a squishy thing and live to tell about it.

PETER: On a previous trip to South Africa, my guide, Jama, and I went to Soweto one day armed with a gift and a wish. Our gift was a pound of dried mopane worms; our wish was that our hosts, a family Jama knew, would prepare their favorite worm recipe for us. Catherine Lemekwana and her daughter Maria thought we were strange, but happily complied. In their tiny kitchen, they made mopane worm casserole with onions, tomatoes, green peppers, and curry. After Maria washed the worms in the backyard, Catherine cooked them. It was an elaborate

process—boiling the worms for 30 minutes, rinsing them twice, draining them, and then sauteeing them for almost an hour with chopped onions, green pepper, tomatoes, curry, and beef stock. The result reminded me of a tofu dish—good, but without much flavor contribution from what was supposedly the main ingredient.

By contrast, Eric's mother's recipe is much simpler. After soaking the worms, she cuts up a tomato and cooks the worms, tomato, and a pinch of salt in corn oil, adding a sprinkle of curry powder at the end. Eaten with *mielie pap*, the ubiquitous cornmeal porridge, the stew makes an excellent breakfast. Once again, though, the mopane worms are a vehicle for other flavors—innocuous shots of invertebrate protein.

PETER: A mile from Eric's village, down a rutted and rocky dirt track, is Masetoni, a small village on a sloping plain overlooking a bend in the Levhuvhu River. The morning we show up with Eric, a cattle round-up is in progress, with sleek cows and calves being corralled and sprayed for ticks. Eric and his wife Peggy know everyone in Masetoni; it takes only a few minutes for Peggy to round up eight kids who are eager to show us how to catch grasshoppers. Meanwhile, Eric asks a group of teenage boys about edible stink bugs. Although they cheerfully go into a small stand of bushes and emerge with a few examples, it isn't really stink bug season. Grasshoppers are better now.

The kids break off branches from small bushes and then fan out to rake and swat the grass with them (**above**). When they see something move, they pounce on it, cat-and-mouse style.

FAITH: Many Vendan names have specific meanings that say something important to the family or others, or protect the child or the family. In the latter category are two of our grasshopper-hunting party, Tshaveheni (Be-Afraid-of-Me), 11, and Thivhashavhi (I'm-Not-Afraid-of-You), 9. In the former category are Aluwani (Grow-Up), 8; Mulalo (Peace), 10; and Hulisani (Respect), 4. Three of the other kids have names like Brenda, which don't have specific meanings, so the generalization is imperfect.

All of the kids are shy at first but warm up as they concentrate on the task at hand. When they proudly present us with their haul (**top right**), I ask what happens to the bugs after they're caught.

Be-Afraid-of-Me: We take them home and cook them and eat them with porridge.

F: How often do you go out to catch insects?

I'm-Not-Afraid-of-You: When we want to or when there's nothing else to eat. Maybe one or five times a week.

F: So you really like these insects?

Peace: They're better than meat.

Not all of the children agree with him. A general discussion of the relative culinary merits of insect species ensues. Brenda likes meat more, she says, though the locusts are nice, too. Peace likes mopane caterpillars more than *muduhwi* ones (the *muduhwi* is a different tree). They eat different types of leaves, he explains, and how good the caterpillar tastes depends on the leaf they eat. Be-Afraid-of-Me likes the *muduhwi*, but not as much as locusts. I'm-Not-Afraid-of-You likes the winged termites most of all. "They're nicer-tasting than the locusts," he says thoughtfully.

In a way, termites end up being my favorite, too. Not because of the taste, but because I liked the women in Eric's village, whom we came to call the Termite Club (**following pages**). Wearing matching gingham aprons, they perch atop a six-foot termite hill, whacking at it with pickaxes. When they knock open a hole, they insert a reed. Soldier termites attack the reed—a bad move, because the women extract the reed and squeegee off the termites into a bowl. The soldiers are washed, fried in oil, and eaten with cornmeal porridge (**left**). The real treat, I decide, is not the soldier termite but I'm-Not-Afraid-of-You's favorite: the winged sexual termite—toasted. I'm sitting there, comparing the textures of the two termite castes, when it occurs to me that I'm having a connoisseur-type experience about members of the insect world.

FAITH: While walking through Vendan villages it's good manners to ask permission before passing through a family's courtyard. Because the path in Eric's village passes through each compound, we meet almost everyone, and we get a little sample of neighborhood life. At one home, we find two women, Vinah Rembuluwani Netshipale and Elsie Ntshengedzeni Netshipale, crushing the fruit of the marula tree in a large bowl to make homemade beer. On the first day, they tell us, the beer is sweet and children can drink it. But by the third day the brew has fermented and it's only for adults.

At another house I sit inside a smoky rondavel with a substantial woman named Tshavhungwe, who is beating her porridge viciously with a handmade whisk. She stops to clap some of the dry *mielie mielie* (cornmeal) on her hands, then clasps her hands prayer-style with the whisk between them, quickly rolling it back and forth. As she works, the porridge begins to thicken. Her breathing thickens, too, and gets louder. Her large size and the heat and smoke inside this small windowless building are causing her problems. Finally, after considerable effort, the porridge is thick enough. She swirls it

and snazzy cars. We see no commercial toys in the village—just lots and lots of ingenuity.

PETER: We spend several days in Eric's village, sleeping at a campground in nearby Kruger National Park. Close to the campground we see herds of impalas, several elephants, and at dawn one morning, a sleepy pair of lions. That night I explain to Faith that it would be better to sleep out under the stars rather than setting up the tent. Three reasons: (1) cooler, (2) less work, (3) not many mosquitos (we were taking anti-malaria pills). She reminds me that we have seen lions only a half-mile away. I show her the nearby game fence. It doesn't look high or strong enough to keep out a hungry lion, but it is a beautiful night with a full moon. Besides, I say, there is another reason not to use the tent: Faith is on the side of the tarp closest to the fence, so she will probably be attacked first, which would give me time to get my cameras and shoot dramatic, lucrative photographs of the battle. Inexplicably, this does not amuse her. I wait until we are driving back to Johannesburg to tell her about how hyenas rip the faces off many sleeping campers each year because that is the part sticking out of the sleeping bag—and how it's recommended that people who don't sleep in tents sleep with chairs over their heads to give them time to fight off the hyenas when they attack. She's appalled.

neatly into a pan, where it will congeal as it cools into the bland milky-white staple food that families eat at every meal. Sweat is sliding down her face in sheets. It's incredible to think that she'll go through this entire process again in the evening —indeed, that she does it at least twice a day, every day.

All the time we are peeking at this village, the village is peeking at us. We are followed around the village by thirteen-year-old Azwifarwi,

steering his homemade Mercedes (**above**). His is one of several handsome do-it-yourself vehicles made from bits of wire, pieces of plastic, wood, carved and "found" wheels, and foam rubber that can be steered with a working steering wheel. Azwifarwi (his name means Don't-Touch) built this car himself and has built others for his brother. Other boys here have built buses, vans,

Uganda

In the lush forests of the Ssese Islands, a small archipelago in Lake Victoria, a village farmer hunts for dead palm trees, a source of palm grubs. *Inset:* Stirred with a palm leaf stem, palm grubs are sautéed in their own oil over a fire.

Uganda

January

PETER: We land at Entebbe airport, catch a bus, then rattle along the two-lane blacktop highway to Kampala past crumbling colonial architecture and billboards about AIDS and condoms. In 1976, it was in Entebbe that Israeli paratroopers stormed an Air France jet after a week-long standoff with Palestinian hijackers. Back then, Uganda was run by big Idi Amin, who killed 300,000 people and stole so much money that the word "kleptocracy" was invented to describe his rule. He fled to Saudi Arabia in 1979 just before neighboring Tanzania invaded the country to get rid of him. Now the country finally has a half-decent government and maybe the fastest-growing economy on the continent.

Not that everything is terrific. The big businesses have armed guards, murderous rebels prowl the north, and the country has one of the worst AIDS epidemics in the world. Still, it is unbelievably lush and green. When I see the chain of Ssese Islands rising from huge, majestic Lake Victoria, I understand why Churchill called Uganda "the pearl of Africa." Unfortunately, Lake Victoria is clogged by floating mats of invading water hyacinth and the water is unswimmable due to *bilharzia*, waterborne parasitic blood flukes. So much for the pearl.

FAITH: We've been told that Ugandans eat termites, lake flies, grasshoppers, and *masinya* (palm worms). This turns out not to be entirely true. Many people eat the other insects, but only the few Ugandans on the Ssese

Islands eat *masinya*. Its host plant, the moriche palm, is abundant there. Joseph Kawunde, 56, a former Ssese Islander, tells us this when we meet in his adopted village of Bweyogerere, an hour northwest of Kampala. When we show him grubs a neighbor has gathered, he turns up his nose. Wrong grubs, says Joseph. "The ones we eat don't have legs. These grubs resemble *masinya* like we resemble monkeys."

Everyone laughs, including the neighbors who have gathered. "If I find you eating these things," Joseph warns us, "I won't talk with you anymore and I won't respect you." He picks up one of the ersatz grubs and flings it away. "It's like an old cow," he scoffs, playing to his audience. A showman!

The grubs he bad-mouths are eaten in other parts of Africa but not here. Ultimately, Joseph decides to help us find real *masinya*. We and most of the village follow him into the forest. Between brandishing his axe and joking with the crowd, he cuts into a fallen palm tree (**right**) and pulls out the coveted white grubs. After extracting a bowl of *masinya*, he carries them home and prepares to cook them under the curious gaze of his neighbors.

Joseph removes their intestines, then sautés the palm worms with salt, curry powder, and yellow onions from his garden (**following pages**). He tells us that when the onions are browned, the grubs are cooked. The neigh-

bors sample the grubs and describe them as "interesting." Some children come back for seconds.

When we come to the Ssese Islands, we see that Joseph's love of *masinya* is commonplace here. In the early morning on the island of Bugala we walk through a driving rain to the house of Sowedi Kanuhamda, a lumber trader who is going to hunt palm worms with his neighbor, James Dyekwaso. Together we tromp into the heavy forest where the dark scent of percolating earth mingles with the sweet smell of vegetation. Wet to the bone, we slog along single file on a narrow trail, through knee-deep water and muddy runoff (**previous pages**). When they find decaying palm trees, Sowedi, 50, and James, 23 (**left**), take turns with the axe, making chunks of hard wet palm wood and globs of squishy fermenting pulp fly.

As they gather up the worms, James recalls for me the first time he saw his father make palm grub soup: "Ahhh, it was very delicious! I told him, 'I will give you a whole chicken if you give me seven more *masinya*.'" When James began to go out to look for them himself, he says, "I told my father, 'You have done a very bad thing to me. I won't eat any food other than my *masinya*.'"

He keeps telling me that he wouldn't eat anything but palm grubs—that he is addicted—so I ask him what he'd do without *masinya*. Palm worms aren't always in season, after all.

"I'll starve," he said.

"What if I steal your *masinya*?"

"I will mourn my *masinya* and I'll curse you!"

James really likes palm grubs.

FAITH: Children in Bweyogerere village don't eat palm grubs, but they do eat a lot of termites. They show us their technique for termite capture: First, hack into a waist-high termite mound to expose tunnels; second, cover the tunnel entrances with a cloth; third, wait while soldier termites attack the invading cloth; fourth, yank away cloth, pick off insects, and eat them (**top left** and **above**). The kids are excited by the heat of the battle. "You must wait. They'll bite you!" "Oh, that one's too small. I don't want it!" "They're finished! They're finished!" "Where's the other hole?" "Here—you dig this one." "But the queen's there!" "Not there!" Peter eats the termites with the kids. I

don't, explaining to anyone listening (no one is) that I won't eat a live cow and shouldn't be expected to eat live termites.

PETER: In Uganda, snacking on termites is like raiding the refrigerator in the U.S., except that raiding a termite mound is more work. The kids grab the soldiers by the rear because the soldiers have clamped onto the cloth with their jaws. Once they detach the termites, like pulling burrs off a shirt, the kids bite off the heads. Not bad—crunchy and nutty— but the bites are too little to get a fix on the taste.

This snack is not for the squeamish. The soldiers squirt a defensive goo from

Termite Farming

Macrotermes bellicosus, a commonly eaten African termite, uses its excrement to construct spaces for cultivating edible fungus—a delicacy. Among the more commonly cultivated fungi are species in the genus *Termitomyces*, found only in termite colonies. One colony can be more than 90 feet wide and can have two million individuals. Most termite-eaters live in areas where the staple food is maize, which lacks two essential amino acids found in termites. Also, soil from termite mounds is eaten by pregnant women; the nutrient-rich loam is an excellent dietary supplement.

their armored heads when attacked and have enlarged biting jaws. In fact, soldiers' jaws are so specialized for attack that soldiers aren't capable of foraging—worker termites feed them with regurgitation or anal secretions. But the termites are helpless in the fight against these kids. The only casualty is the shirt they put over the hole, which gets stained with the brown defensive goo.

FAITH: We take our exploration back into Kampala to the enormous Shauriyako Market. Shauriyako means "look after yourself," so we do, searching the jam-packed stalls and navigating around the semiautomatic-toting guardsmen strolling

about. We like the prices on the food, car parts, and shaman charms, but the cleaner, costlier Nakasero Market, near the bus station, is where we find insects for sale. In Stall No. 70, Kironde Edward sells *mpawu* (dried termites); nearby, in Stall No. 68, Moses Kasule and his mother, Prossy (**lower left**), offer grasshoppers. When I ask where he gets them, Moses says earnestly, "I have sources in the fields who collect these insects." Prossy offers me a taste of her salty dried grasshoppers and I can't refuse. "They're nice," I tell her, which isn't completely untrue.

Peru

Two young Inca herders smile shyly at the camera before disappearing down the mountainside to tend their llamas and sheep in the high mountain passes above the city of Cuzco, on the southern slopes of the Andes. *Inset:* Dry-roasted parch corn and *waykjuiro* worms.

Peru

February

FAITH: Every turn we make on this jagged mountain road is taking a year off my life, and even the straight parts are not exactly reassuring—too many wooden crosses marking the spots where cars and trucks have tumbled off the cliffs. It's a tight squeeze passing the approaching trucks that flag us down, wanting to know whether we've passed police checkpoints for coca leaves. We're at 10,000 feet in the Andes Mountains in south central Peru. The views are crowded with jagged mountain peaks, ancient Inca terraces, herds of llamas, adobe-brick villages, and clouds of mist. Everything except guardrails.

Ángel Callañaupa, artist and backwoodsman, and Raúl Jaimes, Ángel's nephew and our translator, are taking us to meet their cousins in the foothills of the snow-capped Chicón mountain. The nearest settlement

of any size is the town of Urubamba. Houses in Urubamba are all made of adobe, though I can't see much of the adobe because of the hand-painted posters and billboards plastered on every vertical surface— political slogans and notices from the mayor. Three-wheeled red scooter-taxis totter over cobbled streets; oxen and cattle amble by fruit stands; Andinos in wool hats huddle together selling flat loaves of bread. It's windy and cold, even though February is summer in South America.

We cross over a flooding stream into the sparsely populated village of Chicón and drive until the runoff is too deep for our van. Then we start walking. On either side of the valley are farmhouses surrounded by fields of corn and potatoes. Beyond the fields are the Andes, towering over everything. After twenty minutes we arrive at the farmyard belonging to Ángel's cousin, Nicasio

Huaman, three adobe structures around a muddy courtyard in which the pigs are holding court. Nicasio and his wife Asensia welcome us, and our entomophagy expedition begins within minutes. Before we have time to blink, Ángel is charging up the valley wall, Nicasio in tow, yelling to us in Spanish and his native Quechua about the *tayno kuro*. Quechua is the language of the Quechua people who created the powerful Inca empire before the Spanish conquest and whose descendants still live here. *Tayno* is the name of the worm that Nicasio gathers from the *arawanku* plant and *kuro* means worm.

When we catch up with Ángel and Nicasio—not easy at this altitude where every step feels like ten—they are cutting the dried stalks from the center of the *arawanku* plant (**below**). Nicasio's 9-year-old nephew, Edelberto, grabs the torchlike stalks, which shower white

cottony material, puts them on the ground, and splits them lengthwise with the machete. A half-dozen finger-long white worms, the *tayno kuro*, wriggle out of the woody pockets they've eaten in the pulp.

Back in the kitchen house of Nicasio and Asensia's compound, we are conducted to the only seat in the room, a long bench—an honored placement that elevates us slightly above the dirt floor where everyone else sits. Nine fluffy guinea pigs run freely around, staying close to the warm fire. They're not pets: Guinea pigs are an Incan delicacy, eaten at birthdays and other special occasions. I'd like to talk with Asensia about the

traditional Quechua weaving that covers our bench, but she and her 11-year-old daughter Carmen are busy cooking the worms in a traditional Quechua clay pot called a *k'aalla* (**above**).

Asensia says her family always eats worms with parch corn. This is corn that dries completely on the stalk before harvesting. It's heated on the fire until its kernels plump up slightly. This makes a nutritionally sound combination: Corn and worms each lack essential amino acids, but together they provide a balanced meal. Balanced, I'm thinking, but not especially tasty. The corn is okay, but the worms taste like the charred part of a well-done hot dog.

There's no mustard to disguise it, either. By now I've overcome (most of) my revulsion to the idea of eating worms; my lack of enthusiasm isn't sheer cultural prejudice. I just don't think these worms are particularly good. But I have to note that once again I'm in the minority: everyone else (including Peter, of course) thinks the worms are a tasty snack. The family eats them three or four times a week.

PETER: Conquistador architecture, Inca ruins, landscapes at an altitude so high they're literally breathtaking—southern Peru is one of my favorite places on

earth. In the city of Cuzco, the gateway to the mountains, the tourist-to-native ratio is beginning to get out of control, but with an early start you can be way up in the Andes by morning—a time when the poverty of the villages is cloaked by a romantic mist. Even the manure-filled mire of Nicasio's courtyard appears blessed—a fat sow nursing newborns glows pink in the first morning light.

Later, still ruddy from our high-altitude worm hunt, we settle into the family's kitchen to find a shaft of light spearing in through the smoke hole in the roof—illumination straight out of a Rembrandt painting. Guinea pigs are running

around on the muddy floor of their kitchen, cleaning up the food scraps. Supposedly, *cuy*, or guinea pig in Quechua, is delicious. This is my third time in Peru, and Faith can't believe that despite my inclination to eat whatever the locals eat I haven't sampled *cuy*. Neither can I. Meanwhile, the crispy worms with hot kernels of corn make a great snack, reminding me of seared sausages and tender corn nuts. And for dessert, there are strawberries and pears from the garden and the orchard.

FAITH: Ocra Katunki—I'm mesmerized both by the discordant name of this village and by the dried hawk hanging over the window of the house where we've stopped. Our guide, Ángel, has taken us here, three hours due south of Chicón, on the usual appalling Andean roads. Fredi Mollo Cruz, 12, limps over, unsmiling. Then his friend Ángel jumps out and he breaks into a grin. In Quechua, Ángel tells the boy that we want to see the worms that he and his father, Macario, collect—the *waykjuiro*.

Macario and Fredi walk with us: soft rain and bright sun, bald mountaintops, young cornstalks, bearded billy goats. Fredi is limping because he was bitten in the leg by a dog last week. I mention the possibility of seeing a doctor—dog bites are dangerous in areas without rabies shots—but Fredi is dismissive. The *waykjuiro*

eat the *tayanca* tree. Fredi shows us *tayancas* and the worms nestled in their wood. Surprisingly, they're not ugly: two inches long and bright orange with dark spots (**left**). Ángel drops one on Fredi's cap (**above**), which makes the boy laugh. Macario, largely silent thus far, tells me that they cook the worms with oil and either green or red chiles. How does he choose which chiles? "If we want to cry," he says, "we eat the red ones."

PETER: From Ocra Katunki we wind down a dirt road to the nearby village of Chinchapujio, which instantly gives me a bad feeling. Passersby greet smiles with blank stares. The town

itself is unpleasantly raw: All the streets are crisscrossed with sewer ditch construction, the old church is collapsing, the new one is guarded by two big, ferocious dogs and a gate, and the new guest house looks like someone put a junior high school and a Motel 6 in a blender. Worse, it's locked and no one can find the guy with the key. At least it's cheap. Four of us can stay in two rooms for a total of 10 *soles*—less than $4 U.S. We have a supper of beer, cheese, and bread in a small dirt-floored restaurant while we wait for the key. When it finally arrives, I'm not surprised to see that the rooms are filthy and that the toilet is a nearby vacant lot. We

bunk down early (no lights, so no choice, actually). Although Selso (our driver) sleeps in his van to guard it, kids let the air out of his tires in the night.

The first family we visit the next morning, the Ochoas, collect *waykjuiro* from infested tree branches and roast the worms directly on the embers of the fire (**top left**). Bernadina Ochoa is cooking a vegetable peel stew for the pigs in one pot, and the family's vegetable soup in another. She lets them both simmer while she roasts the worms for a breakfast treat. To judge by how quickly everyone eats them, it isn't a bad way to

start the day.

We then visit Salvador Ticona and his family. The fuzzy *waytampu* caterpillars aren't edible until the pupal stage, so Salvador (**following page**) keeps the larvae on the trees in his courtyard until they mature. A caterpillar ranch! As Salvador's daughter fries *waykjuiro*, this wry, thoughtful man jokes with us with a cheer that makes Chinchapujio less depressing.

FAITH: " *Waykjuiro*," Salvador says, "need to be *crunchy*." He reaches into the pan and takes a bite to test its doneness. I'm determined to try one hot and first, not cold and last, as I usually do. When Salvador proclaims

that the worms are perfect, I grab a small one that's just about an inch long and put the whole thing in my mouth. It's steaming hot, it tastes like the earth, and it has a stronger flavor than, say, palm worms. This surprises me because palm worms have more fat, which usually gives food its distinctive flavor. Maybe I think this because the bacony taste of the palm worms is more familiar, and I can't compare these orange worms to anything else I've eaten. In any case, *waykjuiro* are better than other worms I've eaten in Peru, though quite honestly, this isn't saying much.

PETER: At 4:00 the next morning, our equipment in dry bags, we begin the trip from Cuzco in the Andes to the rain forest in the Alta Urubamba River region. By dawn we are winding our way up to a 13,000-foot pass on a dirt road (**following pages**). We see ranches and stone corrals full of llamas at nearly every switchback. At the freezing summit the three men leave offerings and a prayer at a crude stone chapel. Either the prayer or the offerings aren't accepted, because twenty minutes later—*wham!*—a van suspension bar snaps. Ten hours from our destination, we're stranded. After makeshift repairs, we limp into the nearest town, Amparares—no phone, and no mechanic. However, it does have a restaurant with sheep soup for breakfast.

FAITH: In Amparares we get good news—a truck that takes passengers is leaving shortly and going in our direction. The bad news is, we're packed in with seventy people who have not heeded the one-carry-on limit. There are no seat assignments. Once underway, everything is fine (relatively speaking) except for the frequent spots where tiny mountain streams have swollen into raging rivers that rush over the skinny one-lane road and down the mountainside. Sometimes we have to ford these things while somehow passing other trucks. The road, just bulldozed a few years ago, is only about eight feet wide. The other trucks are loaded with passengers, too, and it's carnival time, so the people in villages along the road pitch celebratory buckets of water at us. The real fun comes when the driver

slows, Raúl shouts for us to take cover, and we pass through a huge waterfall that is smashing onto the road from an overhanging cliff. Everyone is laughing and takes it in stride.

When the driver stops for lunch in Colca, we find another ride—a bus, complete with actual seats and a radio blaring Peruvian pop. After switching to a different bus and then a van, we arrive in darkness 13 hours later at the hamlet of Koribeni—home to a bus stop, two tiny stores that sell beer and cold Coke (Heaven!), and a couple hundred people in spare forest dwellings, some of whom are cousins of Raúl and Ángel.

PETER: We crawl out of our tent after a predawn cloudburst and it's finally light enough to see the tin-roofed bamboo home of Daniel Piña Real, 30, and his family—the cousins. We're

camped at the edge of his garden, near the path to the river where they bathe and get water. Sitting on plank benches, we have a breakfast of cheese, crackers, coffee, sour oranges, and sweet slimy cacao beans.

Later everyone tromps through a stream, looking under rocks for creepy bugs called *chanchu chanchu* (**above**). The larvae of fish-flies, they look like *zaza-mushi* to me (see pages 32–33). We eat the *chanchu chanchu* alive (Faith doesn't), after pulling off the heads. They taste like fishy rubber, which may be why the Japanese cook their *zaza-mushi* in so much sugar and soy sauce.

FAITH: I talk with Daniel's great-uncle, Don Ignacio, 81, who lives just over the hill. Since the death of his wife a year ago, he has been taking meals with Daniel's family. His wife, Don Ignacio explains, isn't buried yet. "She's in a box by the side of

the house," he says, biting into a sour orange. He tells me no more about this. Instead, he begins talking about eating his pet monkeys and I forget to return to the question of why his dead wife is stored in a box next to his house.

"The fish were plentiful in the early days," he says. "So were monkeys. The small monkeys were best for eating but I would keep a monkey until it got old, then eat it."

"Do you still eat monkey?" I ask, a bit sickened. "Not any more," he says. "People chased them away. We taught them to pick up the wood piles. They could do anything they were taught, but we had to be careful. I had a problem one time with a monkey that liked my gun. Another monkey would start fires on the roof of my restaurant." "Do you miss the monkeys?" I ask. "I miss eating them," Don Ignacio says.

PETER: Later we hike to an overgrown graveyard by the Yanatile River. Daniel chops into the hardwood stump of a fallen *pansona* tree, then his daughter, Marleni, 16, and son, Ramiro, 14, poke through the rotten wood for lunch: *chiro* worms (**top left**). These long, white, hammer-headed beetle larvae, which can measure six inches in length, look like pumped-up versions of the *ulat kayu* (see page 75) in Irian Jaya.

Because Daniel's wife is away at a UNICEF health class, Marleni makes lunch. After building a fire in their earthen stove, she puts the live unwashed worms into a frying pan with vegetable oil (**bottom left**). The *chiro* worms squirm in the hot, spitting oil, but not for long. Soon they are little brown brat-wursts. Everyone in the family loves them, including 4-year-old Daniel (**right**). Marleni says the taste can't be compared to chicken, *cuy,* beef, or fish. I think they taste like pork sausages with crunchy heads.

Daniel shows us how to prepare them *wanta*-style. He wraps the worms in a banana leaf, adds a pinch of salt, and props the tied bundle by the fire. Cooked this way, the worms are more tender, but they look alive and so are less popular with the kids.

Size Matters

Chiro worms are the larvae of longhorn beetles, so called for the amazingly long antennae that characterize the family Cerambycidae. Males of nocturnal species in the family use their long antennae to sweep the bark surface in the hope of touching a female. The wider the sweep, the greater the chances of finding a mate.

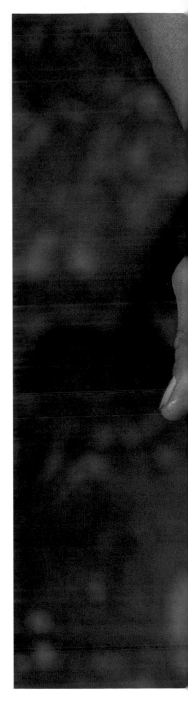

FAITH: We're about twenty hours north of Cuzco in the Peruvian rain forest, jammed with fifty people into a fifty-foot-long motorized longboat. It would fit thirty-five people comfortably. Sitting on the prow, our navigator directs the pilot through the hellishly swift current of the Alta Urubamba River. We survive and are dropped at the village of Yaneriato, where we meet Solomon Mintiani, Lucía Tenteyo, and their family. Next to the family compound, there's a fast-moving stream where they wash dishes and bathe. Solomon says that everyone else in the village uses that stream, too, so he plans to move his house to get more privacy. He and his family are Machigüenga Indians. Our guide, Ángel, who lived with them for a time, tells us that they are good at living off the fruits of the land and that they often eat insects. That's why we're here.

PETER: The first thing they show us is a *kisa-kisa*, a bright green caterpillar with a red and-white racing stripe and tiny fernlike appendages (**top left**). The *kisa-kisa*'s color is a warning: Not all insects are edible. The fern-like appendages are covered with a toxic substance and the bright coloring is an alert to predators: Stay away.

Soon after this defensive signal, we're on the offense, hiking up the valley wall in search of palm grubs. We follow Solomon and his kids to a rotting palm above one of the family's slash-and-burn gardens. The palms here are smaller than those in Indonesia or Uganda, but the grubs are the same size. The real difference is the way they are eaten—uncooked. Raw, raw, raw—that's the spirit! But not for me. After

we return to Solomon's house, the bowl of grubs sits out in the sun for about four hours. By the time people eat them (**bottom left**), the white, wiggling worms are no longer white and no longer wiggling. I photograph everyone else sucking out the insides but I pass.

A late-afternoon foray by Maximo, Solomon's brother, yields a second movable feast (**above**). His wife, Florinda, cooks the bugs for us. When I find out that she has some

beer for sale stashed in their sleeping hut, I buy everyone a warm brew. We sit under a tree until dark, enjoying our conversation and the insect and alcohol buzz.

FAITH: This is the only place we've been where Peter hasn't said he wants to eat the whole bowl of insects. As I've said before, I draw the line at eating live bugs, but there are plenty of other things here keeping me mentally engaged. Lucia weaves in the traditional style, and Solomon is concerned that his people are losing their culture. I'm not sure what Lucia thinks about this because she is incredibly shy with us; I would need more time to earn her trust. Solomon teaches his children about their heritage, has them wear their traditional clothes, and worries that the road the government is constructing behind his house will connect his people to the outside world, ultimately destroying their way of life. Solomon's brother Maximo, a former tribal chief, is less fatalistic but no less concerned. "We have to be careful to maintain what makes us who we are," he says. Solomon is particularly worried about the influx of religious missionaries of all flavors. "He keeps the religion of the Incas—we have no religion except for the sun," he tells us. But his mother has become an evangelical Protestant. And even his brother, the former tribal chief, is Roman Catholic.

Following pages:
Children cross a suspension bridge over the Yanatile River in the town of Santiago, northeast of Cuzco.

Venezuela

The narrow prow of a *bongo*—a 30-foot-long dugout canoe—
pushes up the Orinoco River and deep into the rain forest home
of the Yanomami in southeast Venezuela. *Inset:* Fire-roasted
Theraphosa leblondi, the world's biggest tarantula.

Venezuela

February

FAITH: In 1991, Venezuela turned part of Amazonas, its southernmost state, into a biosphere reserve to protect the land and its people from increasingly rapid ecological degradation. Nothing and no one moves in or out without a permit, theoretically preserving the 20-million-acre area for study. Permits take weeks to arrange; the terms are enforced by the Guardia Nacional, the Venezuelan army, which has posts along the rivers that serve as highways in the rain forest. The biggest highway is the Orinoco River, whose headwaters in northwestern Brazil and southern Venezuela are home to about 23,000 Yanomami. For decades anthropologists have quarreled about whether or not the Yanomami are among the most warlike people on earth. Either way, the Yanomami in Venezuela are the ones we've come to see.

PETER: From Puerto Ayacucho, the capital of Amazonas, we pack our equipment into a bush Cessna and take off into ominous skies. We spend two hours bouncing from one squall to the next in the hands of our pilot, a pudgy kid no more than twenty years old, who deftly flies 250 miles through zero-visibility weather to the tiny grass airstrip at Tamatama, a missionary outpost along the Orinoco. This is the Venezuelan base camp of the New Tribes Mission, a worldwide evangelical group of Protestants that came here in the fifties.

Our Spanish-speaking guide, Javier, has traveled through this area on anthropological expeditions and learned the language of the Yanomami. Carlos, Faith's interpreter for Spanish to English, is a Costa Rican ex-Sandinista guerrilla who's become a backwoods guide since the revolution. In Tamatama, Carlos arranges for his friend Santos, who has a motorized dugout canoe, or *bongo*, to take us to our destination: Sejal, a Yanomami village less than an hour down the river.

The *bongo* is too small for everyone to fit in, so Santos takes me, Javier, the equipment, and all the food on the first trip. We offload at a rocky outcrop where kids scramble about (**right**). Then Santos returns upriver to get Faith and Carlos, Javier goes off to find the village chief, and I sit watching our stuff as several men feel the outside of the bags, trying to guess what's inside. "Machetes?" one asks in Spanish, feeling my tripod bag. It appears they've had visitors before.

An hour later the chief has conducted us to the new visitors' hut, which is adjacent to a slash-and-burn garden (**below**). While I hang the hammocks and wait for Carlos and Faith to arrive, Javier, sweating profusely in a thick shirt that protects

him from malarial mosquitos, begins making lunch: coleslaw with mayonnaise, raisins, and canned pineapple. Twenty villagers gather around to see the strange food. We are a sideshow.

As if in a bad comedy, the scene is interrupted by the arrival of a teenage Guardia soldier in camouflage fatigues with a rifle. He orders us to pack everything up and return to Tamatama. We fight with every weapon at our disposal: reason, the permit, even an invitation to lunch. No way. Back to the *bongo*. Javier is really sweating now. Our stuff is in the boat so there's no room for the soldier. We offload bags. All the villagers who are so disappointed at our depar-

ture now seem less unhappy. If we don't come back, that's okay—we've left a week's food on the riverbank.

FAITH: While we wait for the *bongo*'s return, Carlos and I walk around the mission, which is almost all there is to Tamatama. We end our walk at the adjacent Guardia outpost, where an unsmiling soldier motions with his rifle to take a seat outside if we like. Carlos chats with the guy, and before too long we're looking at his boot camp photos. I watch the Orinoco, thankful for the cooling canopy overhead, and idly listen to the Darth Vaderish voice speaking on the police radio inside. My Spanish is poor,

but it requires no great acumen to see that Carlos is now listening intently to the voices bellowing through the concrete walls. Darth Vader is upset—shouting at someone in this post about the village we're heading to. When Carlos says we're in trouble, I feign surprise (which doesn't amuse him). We learn that Javier should have stopped at this outpost when the *bongo* passed by and didn't. Darth Vader is now telling the post here to get our *bongo* back.

PETER: Heading back, the young soldier ties his life jacket to his rifle, not himself. As we approach Tamatama,

Faith and Carlos come down to the riverbank from the Guardia post. She's annoyed. "Don't speak Spanish," she says. "If they don't know that you can understand them, maybe you'll find out if we're ever going to get out of here."

FAITH: At the outpost, we struggle to break free of the red tape and fail. We're told to stay overnight at the *bongo* owner's house and then go to the main Guardia post in La Esmeralda, two hours upriver. To do this, we have to borrow a bigger boat and fuel. We string up our hammocks in an outbuilding and try to sleep. Since everyone's hammock is connected to the same central pole, any movement by one is felt by

all, like puppets on a string. What fun.

PETER: The next morning we motor in the borrowed *bongo* to Esmeralda, where we have to sit around for an hour before we get permission to go where we already have permission to go. We scrounge gasoline and Carlos buys a smoked fish, then we shove off downriver. The wind picks up and the sky looks like it wants to pick a fight. By the time we land back at Tamatama to pick up our stuff, it's raining hard enough for a strong fish to swim into the clouds. But we're finally "legal" in the eyes of the men with the guns.

FAITH: I'm incredibly curious as we approach the riverbank at Sejal. I've read more about the Yanomami than about any of the other indigenous peoples we've met, largely because so many anthropologists have come to such conflicting conclusions about them. But it's hard to notice much because Carlos keeps nattering on about the world according to Carlos. We drop our stuff in the visitors' house, introduce ourselves to the second-in-charge (the headman's gone hunting) and take a quick look around before heading into the forest. Except for the gifts we've brought, we don't seem to interest the Yanomami here. They live on the fringes of Yanomami culture both literally and figuratively, and are used to visitors.

PETER: Thirty-six hours after being hauled back to Tamatama, we're again in the Sejal visitors' hut. Everything is as before; we even have our food back. Javier organizes a rain forest recon to see what kind of bugs are out there, with the emphasis on giant tarantulas. Most of the men I met the first time have gone hunting for several days, so Chaurino Perez Andrate, 17, Gregorio Lopez, 16, and some younger boys take us into the forest.

It turns out that these young hunters and gatherers aren't much good at hunting and gathering. They're not terribly interested in insects and stop every few hundred yards to shoot long arrows at birds high in the trees (**top right**). But they never hit anything, so sometimes the arrows don't come down and they have to climb the trees to retrieve them. Lousy shots, but good climbers. Maybe this is why they were left behind by the hunting party.

Eventually they cut down a tree containing a termite

nest (**top left**) but abandon it, complaining that the termites are the wrong size. They also chop open old palm logs and find a cache of palm worms like we have seen in Peru, Uganda, and Indonesia. Gregorio transports the worms back to the village by wrapping them in palm leaves (**bottom, left**).

Tarantulas are less work. After locating a spider's burrow, the kids strip the bark from a long thin vine (vines hang on almost every tree), tie a knot in one end, and feed that end into the hole while twirling the vine between the palms of their hands. When a tarantula grabs the knot, the boys slowly pull the vine out with the spider hanging on—like crabbing. But these tarantulas are no prize: they're not the big kind, and in any case, this group of Yanomami likes faster food.

FAITH: When I remark later about the boys' surprising lack of skill, Santos Perez,

the Yanomami man in charge of visitors, tells me, "They're only learning." But Chaurino tells me later that catching tarantulas and insects is something he rarely does, though some of the older people in the village do. He works in his garden and wants to be a tour guide like Javier and Carlos. His young friends almost all give the same answer when asked whether they hunt insects to eat: "Not really," they say. What I learn is that the Yanomami culture— here, at least—is changing fairly rapidly.

This village of Sejal is a hybrid of old and new. Margarita, the wife of a shaman, follows tradition and won't divulge her Yanomami name to outsiders. She also dresses traditionally (**preceding pages,** Margarita waits for her potatoes to cook), but she and her co-wife Susana weave baskets that they sometimes

exchange for tourist dollars, a cultural departure for the Yanomami, who until recently had no source of, or use for, hard currency. Her grandchildren wear t-shirts with English messages. "The White Sox Rule" and "Buxom Beauties of Mud Wrestling."

At first I'm annoyed that we're not spending time with more traditional Yanomami who haven't been as "touched" by the outside world as have the people here. Then I realize that Sejal exemplifies an utterly contemporary phenomenon: eco-tourism, both preserving and changing a society. Santos tells me that he would gladly exchange parts of his cultural heritage for a life that is less hard, though he maintains, "We keep our culture alive among ourselves." He also says, "We live better now, away from the warring, which was very bad." (The Yanomami have complex traditions that include fierce

fighting and retribution.)

Chaurino is somewhere in the middle of this transition. On the one hand, he has no real interest in traditional fare, like spiders and palm worms, when he can trade produce from his garden for the packaged noodles and rice that come from downstream. On the other hand, his family is adhering to tradition by arranging a future marriage for him to a girl who is now 5 years old. So it goes throughout Sejal. We had to go tarantula-hunting with kids because many of the men who hadn't gone hunting were at the shaman's house, learning to be shamans, and using hollow reeds to blow a hallucinogenic powder called *yoppo* up each other's noses. As we cook our tarantulas in a neighboring house, loud— and utterly authentic— retching noises from the men taking *yoppo* resound throughout the village.

PETER: It is our last full day in Sejal and we have not seen any of the really big tarantulas. *Theraphosa leblondi*, the biggest spider in the world, is the size of a dinner plate, but so far the ones we've caught are just the size of a saucer. On a final hunt, hour after hour drips by as we wind our way through the forest, stopping to fish at every tarantula hole. The heat makes everyone edgy and slow. Several spiders escape into their burrows as the boys gawk. Finally Santos hauls up a decent-sized *T. leblondi*. He carries it back to the village on his machete (**bottom**).

Because the boys overcooked the smaller ones on the first day (**top**), I tell them I prefer my tarantulas medium-rare. Chaurino stuns the *leblondi* by whacking it with a stick, gathers its legs, and lowers it onto the fire. The spider makes a final hiss as its insides heat up and it shoots out a yard-long spurt of hot juice. After it is roasted for about seven minutes (**far right,** Chaurino and cooked tarantula), its charred hairs are rubbed away and the legs pulled off. When we crack them open, there's white meat. No goo—this creature has actual muscles. The same with the abdomen, which has the most meat. It's tasty—like smoky crab. I wish we had a half-dozen more.

When we return to Tamatama, we see the Guardia again, but now they're our friends. One guy gives Faith a bullet on a necklace, no doubt a symbol of his power of life and death over her. We give him a pen and a photo business card, symbolizing our power to make him look and sound foolish. Which is mightier?

Actually, I don't bear the

Velcro Warning

Theraphosa leblondi warns its enemies with a hissing sound produced in a unique manner. Parts of its first and second pairs of legs and pedipalps (appendages behind the first pair of claws) are covered with hairs that end in hooks. By a Velcro-like entangling and disentangling of the hooks on one set of limbs from the hairs on the opposing sets, the spider produces a noise startling enough to give it time to escape.

Guardia any malice. I know they have little understanding of why they are being ordered to limit the locals' contact with the outside world, given that the locals themselves seem to want everything they can get from it. As for the Yanomami, I hope they don't become prisoners in their own museum.

United States

PETER & FAITH: According to Larry Peterman (**right,** at his candy store), founder of the HotLix candy company, most Americans have two reactions to eating bugs: disbelief and disgust. In fact, he says, they buy his company's insect-related sweets and snacks *because* they think they're unbelievable and disgusting.

Eating insects has not always been a joke. Humans have a long history of entomophagy. In North America, insects were especially important to diets in the desert Southwest and Mexico, where Native American groups such as the Paiutes feasted for centuries on the pine-feeding caterpillar of the Pandora moth. They smoked them out of the trees by building small brush fires under the canopy, thus gathering and cooking them at the same time. In more recent times, the Paiutes and the U.S. Forest Service fought over spraying forests to control the Pandora moth. The battle ended in victory for the Paiutes—a triumph of food over forestry.

Hopper Whoppers

In an experiment, Utah state archaeologist David Madsen examined the energy efficiency of collecting live Mormon crickets (*Anabrus simplex*), a former Native American food. In some situations, Madsen's group was able to collect crickets at a rate of almost 18½ pounds per hour. At that rate, he says, a cricket collector who works for an hour "accomplishes as much as one collecting 87 chili dogs, 49 slices of pizza, or 43 Big Macs." The Native American penchant for crickets, he concluded, "made a great deal of economic sense."

Perhaps the greatest of the Native American insect gourmands were the Great Basin peoples who lived around the Great Salt Lake in what is now Utah. Every so often huge swarms of grasshoppers are blown into the lake, where they drown. Their salty bodies wash to the shore and dry in the sun. The carpet of insects on the sand can contain as many as 10,000 insects per foot of shoreline. According to Utah state archaeologist David Madsen, who has examined ancient insect fragments and human excrement from lakeside caves, hunter-gatherers harvested and ate the dried grasshoppers for generations. By scooping up insects from the lake shore, Madsen and his colleagues discovered that one individual could collect 200 pounds of sun-dried grasshoppers in an hour—enough food to feed a family for weeks.

And why shouldn't Native Americans have eaten them? On a percentage basis insects usually have about the same protein content as the flesh of mammals and birds and are rich in vitamins and minerals. Dried insects, like the grasshoppers in Utah, have double or triple the protein. (The entomologist May Berenbaum jokes that maybe it's the worm a day and not the apple it's eating that keeps the doctor away.) True, the hard, chitinous exoskeleton of adult insects is mostly indigestible, but so is apple skin; comprising only a small percentage of an insect's mass, the shell doesn't affect its food value. All in all, insects are way up on the nutrition charts.

Entomophagists like to point out that insects convert ingested food to protein more efficiently than traditional food animals do, and

hence are better for the planet. According to University of Wisconsin entomologist Richard L. Lindroth, the Efficiency Conversion Index (the percent of the weight of food eaten by an animal that turns into weight gained) of range-fed beef cattle is about three. The other ninety-seven percent is excreted. In contrast, range-fed caterpillars have an ECI of twenty to thirty, depending on the species. It takes ten times as much food, in other words, to produce a pound of cow as it does to produce a pound of caterpillar. Because some insects, unlike cows, can be fed agricultural waste, it's hard not to concede the bug buffs' point.

Logic like this is why bug fanciers have pushed entomophagy for decades—and have been ridiculed for just as long. As far back as 1939, C.H. Curran, the insect

curator of the American Museum of Natural History, was scoffing at the "people who have suggested that we should eat insects" for "seeking notoriety or being facetious." Not himself squeamish about entomophagy, Curran admitted that many insects were edible. "However," he argued, "it is absurd to urge upon a people blessed with a superabundance of good, delectable food, the advantage of eating something which is likely to prove less agreeable to the palate than the things to which we are now accustomed."

But people keep doing just that. A leading enthusiast, entomologist Florence S. Dunkel, the current editor of *The Food Insects Newsletter,* tells us "Insects are great gourmet delights and certainly the food of the future. Entomophagy is ecologically sound. You can raise all this nutritional power literally in your kitchen cupboard. No need for acres and acres of rangeland!"

Dunkel may be right, but Americans in the past have been cautious about embracing new food: It took a decade for sushi to invade America, but the Colonel polysaturated Japan in only a few years. (Kentucky Fried Chicken is everywhere now in Asia.) On Planet Big Mac we look to the future, eschewing bugs in our software as well as on our dinnerware. Logic won't put insects on the average American's plate. It will take clever marketing. But most importantly, there needs to be an economic incentive before the big food companies will jump in. Peterman's HotLix company is expanding its business because it believes that some acceptance is on the horizon. But he also believes that most Americans will continue to be disgusted by insects for a long time.

PETER: A former Monsanto researcher, HotLix founder Peterman was tentative back in 1991 when he augmented his candy line with his first insect product: tequila-flavored lollipops with a mealworm inside. "You can't just pick up a bug off the ground and put it into a piece of candy," he says. "You have to test it to make sure it is not toxic and to make sure the legs won't fall off when you cook it. You have to get all the 'bugs' worked out of it."

The success of the mealworm candy led to experiments with chocolate-dipped scorpions (**right**), candy-coated scorpions called *InsectNsides* (**far right**), and candied apples with mealworms (*page 192*). Next to Peterman's tiny office, workers put dried mealworms into molds and cover them with hot apple-flavored syrup to make *Worm-in-Apple* suckers (**opposite top**). A similar process produces crème de menthe *Cricket Lick-It* suckers (**following pages**).

The insect business will grow, Peterman believes. "I think there will be room for a gourmet insect-type thing, because insects *do* have different flavors," he says. Some insects, though, "don't lend themselves for mass production, so they would be like a gourmet item—like brains." The Jerusalem cricket is an example. "You've heard of hummingbird tongues? These crickets have a little brain cavity and they do have just a little bit of stuff there... It's probably an acquired taste, like caviar. There might even be insect snobs at one point, like wine snobs."

FAITH: In their second year of business, the Bassett family learned a hard lesson when they accidentally killed their product line. "We sprayed pesticide to get rid of a beetle," says Russ Bassett (**above,** at left with his father Dale), president of the Visalia, California, bug farm

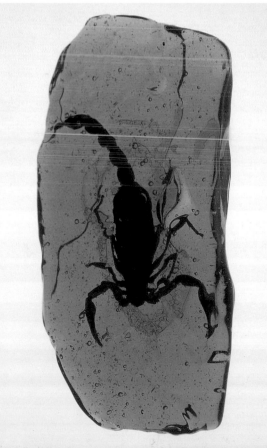

that supplies HotLix with crickets and mealworms. "The beetles survived but the crickets didn't. Now it's taboo to bring any insecticide around here."

Today, Bassett Cricket Ranch is a multi-million-dollar business that sells insects for pet food and fishing bait. Russ's parents opened the ranch in 1969 after his dad tried crickets as bait, liked the results, and began selling the insects from his sporting-goods store. When he couldn't find enough crickets to sell, he began raising them at home. The company grew rapidly. It now sells more than two million live crickets a week at peak season.

Two million crickets a week sounds impressive to me, but I learn differently when I talk with Robert Kok, an agricultural-systems engineer at McGill University. An advocate of insect agriculture, Kok explains that the industrial bug ranches of the future would operate on an entirely

different scale. A world-class insect factory, in his view, should produce 1,800 tons of insect matter a day. One hundred factories, he calculates enthusiastically, could meet the needs of the 600 million people who are now chronically deficient in protein.

People like the Bassetts, Kok admits, will be hard-pressed to survive the advent of large-scale agribusiness. "Tofu is a prime example. A lot of people had small tofu production plants. It became successful. Everybody scaled up. Large supermarkets moved in, and all those little tofu producers died."

Still, Kok recognizes the need to start small; he's done it himself. His own exploration of insect agriculture has been limited to making a few ground-insect hot dogs and feeding them to friends. "Most people say they're no worse than ordinary wieners," he says.

PETER: The man who first got me interested in edible insects is Gene R. DeFoliart, a mild-mannered entomologist at the University of Wisconsin. It was through the *The Food Insects Newsletter*, which Gene founded in 1988, that I came to know entomophagy as something weirdly wonderful, not lethally loathsome. Gene believes that the "age of edible insects may just be dawning," but he preaches his message with good humor and a grain of salt.

Gene retired in 1995, passing on the editorship of the newsletter to Montana entomologist Florence Dunkel, but he retains his interest in entomophagy. Gene thinks that embracing traditional insect foods in developing countries could lead to enhanced forest conservation (to protect edible species); reduced pesticide use; reduced organic pollution through the recycling of agricultural and forestry wastes into high-quality food or animal foodstuffs; and more efficient utilization of resources. All will be a boon for the emerging world, increasing food production and income for poor rural families while decreasing pressure for land clearing, pesticides, and intensive agriculture. A big problem, in his view, is that the attitude of the First World toward insect food could prevent the Third World from returning to this traditional food source. Often progress means abandoning old habits. In this case, Gene argues, progress will come from keeping them.

FAITH: Safiya Carter-Thompson, 12, and two friends, Josh Olson and Alex Baker-Lubin, both 11, (**above,** in foreground) were invited to an insect-cooking lesson by entomologist Leslie Saul of the San Francisco Insect Zoo and her husband Norman Gershenz. On the menu: chocolate-chip mealworm cookies and a cricket frittata. I asked Safiya (**right,** tasting her first insect) what she thought when she first learned she was going to cook and eat insects.

Safiya: It was kind of like, 'eeeww bugs!' But I thought I'd do it because it was something new. My brother was disgusted.

F: Now that you've done it, what do you think of the idea of eating insects?

Safiya: I don't feel so weird about it anymore. If I had the same chance, I'd do it over again, but it really depends on the insect. I couldn't eat just any bug. I don't think I'd eat any that look funny to me like caterpillars. They look too squishy to eat.

F: There are a lot of cultures around the world for which eating insects is perfectly acceptable. This is a very large unexploited source of nutrition that Americans won't eat. Why do you think we're like that in the United States?

Safiya: Americans are kind of picky. When something new comes along, like eating insects, they probably wouldn't try it because they're not used to trying interesting new things.

F: When you went back to school, did you tell all your classmates about this experience?

Safiya: Yes, and Alex brought some of the cookies we made and people could try them. Only a couple of people did. Then Alex ate too many and became sick and had to go home.

F: Do you think it was the cookies that made him sick or the idea of eating them?

Safiya: I think maybe both.

Afterword
Teen flips over roach chips

VENDA, SOUTH AFRICA, was where I had one of *those* moments. I was talking with a group of women we privately called the Termite Club. "Club" because they had showed up to talk to Peter and me in identical red-and-white gingham dresses, "termite" because they were perched on a tall termite mound in those outfits, catching the insects for lunch. They were fishing around in the holes they'd dug, when somehow it came out that I had eaten scorpions in China.

As one, the members of the Termite Club put down their bugs and stared at me in shock, laughing nervously. Equally nervous—there were a lot of termites around, and they bite—I said something about how it is interesting that there are Chinese people who eat things that South Africans won't eat, and people in the U.S. who won't eat what South Africans eat. Everybody's different, I offered, or some other cliché to this effect. The scorpions are fried in oil, like the locusts cooked in South Africa....

I stopped, because my questioners were making gagging noises. "I would not eat them," Patricia Maphangule said.

"The Chinese breed them," I told her, "and they raise them in their homes." Stunned silence. This was too unbelievable, too shocking for words. Shaking their heads, the Termite Club returned to fishing for the little scuttling termite bodies that they would stir-fry with tomato and eat whole for lunch. They kept staring at me, wondering if I was deliberately trying to horrify them.

Working on this book, I got to be familiar with this reaction. In fact, I had to come to terms with it myself.

The only conversations about insects I can remember from my past had to do with the best way to kill them. I certainly didn't talk about the best way to cook them. Indeed, I am one of the most unlikely insect eaters you will ever encounter. Even in adulthood, the monsters of my childhood have been hard to banish, and most of them are insects.

As a 14-year-old in Connecticut, I was asked one evening to babysit the children of my parents' friends. They lived around the corner from us and had recently moved from the South. I hardly knew them, but the children played quietly and I settled in with my book. When the kids asked for a snack, I pointed in the direction of the bag of potato chips their mother had pulled from the cupboard before she left. Earlier in the evening something or other had skittered past my foot, startling me. But this was nothing compared to my hysteria when the little girl pulled a potato chip covered with cockroaches out of the bag. I'd never seen one before. Seeing more than one crawling around—seeing them crawling on *food*—I slapped the chips and bugs from the child's hands and threw them across the room. Looking around I suddenly realized the house was full of roaches—they blended in with the walls. The roaches had obviously traveled with the family from the South and were getting used to their new home. The kids couldn't believe that I was freaked out, and I couldn't believe that they weren't as disgusted as I. Years later I lived in a Southern state overrun by heat- and humidity-loving roaches and learned why the children wouldn't be disgusted. But at the time the roaches horrified me. To this day the smell of potato chips makes me queasy.

When I lived in the South, people ate plates full of buglike food—spicy crawdads. Although the crawdads smelled good, I couldn't bring myself to put one in my mouth. Crawdads are arthropods, just like insects, and I wasn't going to eat anything that was in the same phylum as insects. So how did someone like me manage to eat a tarantula? (A small part of one, anyway.)

I've worked hard never to have a personal relationship—good or bad—with my sources of meat protein. Like almost everyone in the developed world, I've always been removed from the source of my food. I do my hunting and gathering at the supermarket. There are no cows in my yard, just plastic-covered slabs of meat by the grill, lying on Styrofoam plates with absorbent pads soaking up the extra juice (blood). The pork chops I eat do not look like pigs; chickens come to me beheaded and befooted, and sometimes sans bones and skin.

You can imagine what it felt like the first time I went to China, when I was confronted with a steamed chicken in its entirety—the Chinese are not wasteful. Eating this chicken that looked like a chicken was a defining moment in my education as a responsible consumer of meat protein. And coming to terms with what the food I eat really looks like is what ultimately allowed me to eat my first insect. On our next trip to China, I ended up with worm casserole on my plate. What did I do? I thought about sausage as I ate the casserole. Eating worms is a lot less appalling to me than eating spiced animal entrails encased in animal intestine.

Eating insects never got easy, but I learned that it is possible. And I now realize the experience changed my life. Dropping by the local supermarket is not the same for me anymore. Though I've always been stunned by the sheer amount and variety of food available in the United States, the shelves of the supermarket carry only a narrow slice of what the world has to offer, one dictated by the preferences of North Americans like me. Except the small percentage of insect parts inadvertently included in our foods (and allowed there by U.S. law), there are no insects—except, sometimes, lobsters, which are relatives of spiders. I don't really want a plethora of insects to choose from in my supermarket. But now I know there could be.   —FAITH D'ALUISIO

Glossary

Animal: In general, a multicellular organism that, unlike plants, cannot synthesize carbohydrates from simple organic or inorganic materials but must ingest them as food. The Animalia, one of the five classical kingdoms of life, includes such diverse creatures as insects, spiders, sponges, mollusks, crustaceans, fish, and mammals. Animals, unlike other life forms that acquire energy by consuming other organisms, tend to be mobile during part or all of their life cycle.

Ant: Insect with membranous wings and elbowed antennae that lives in nests and displays complex social behavior. Within an ant colony there are usually three castes, or forms, of individuals, each with different responsibilities to the society: queens, males, and workers. Ants display some remarkable behaviors, even forcing ants from other colonies or species to work for them. The queen *Bothriomyrmex decapitans* of Africa, for example, allows herself to be dragged by ants from the genus *Tapinoma* into their nest. She then bites off the head of the *Tapinoma* queen and begins laying her own eggs, which are cared for by the "enslaved" *Tapinoma* workers. About 8,000 species are known. (Family: Formicidae.) See *Honeydew, Soldier, Worker.*

Arachnid: Arthropod with body consisting of two major divisions and, usually, six pairs of appendages: one for holding prey (chelicerae); one for touching, antenna-style (pedipalps); and four for walking. (Insects, by contrast, have three pairs of appendages and three body divisions or regions. Tens of thousands of species of arachnid have been described, of which the most well-known are spiders, scorpions, ticks, and mites. With the exception of some of the mites, arachnids are primarily predaceous and feed on a wide range of prey species.

Most inject digestive fluids into their prey before sucking the liquefied remains into their mouths. (Class: Arachnida.)

Arthropod: Invertebrates with hard exoskeletons and segmented appendages, comprising the largest group in the animal kingdom. The phylum Arthropoda includes insects, spiders, lobsters, crabs, and millipedes as well as many, many more obscure creatures. Scientists have identified about a million species of arthropod and believe millions more exist that have not yet been described.

Bee: Winged, hairy-bodied insect equipped with a tongue-like structure used to suck flower nectar for food. Most are solitary—honeybees are an exception. By flying among flowers, bees transfer pollen between blossoms, enabling many species of plants to reproduce. Many plants, including major food crops, cannot bear seed without bee pollination. About 20,000 species of bee are known. (Superfamily: Apoidea.)

Beetle: Insect with thickened forewings that fold protectively over the body (most insects have four wings, two in front and two behind). Beetles are the largest group of insects. More than 250,000 species are known. (Order: Coleoptera.)

Bioluminescence: Emission of light from an organism, as in fireflies. Caused by an internal chemical reaction, bioluminescence produces light with very little heat.

Brooding: Process in which a parent animal incubates eggs. Some insects, especially bees and wasps, build "brood cells": food-filled spaces in the nest in which young insects develop.

Bug: Technically, an insect in the order Hemiptera, distinguished by wings that are partially leathery and partially membranous and by mouthparts adapted to pierce plant or animal tissues. Examples of the more than 30,000 species of true bug include water bugs and stink bugs. Colloquially, the term is used to refer to any insect or creeping invertebrate.

Caterpillar: Wormlike, often brightly colored, juvenile form of the adult butterfly or moth. Sometimes called, imprecisely, "worms," they have 13 segments in their bodies and six eyes. See *Larva.*

Chrysalis (Chrysalid): See *Pupa.*

Cicada: Medium to large insect with a piercing proboscis (beak) for sucking plant juices and, in males, a pair of vibrating organs at the base of the abdomen that produce a high-pitched, droning sound. Cicadas are sometimes called, imprecisely, "locusts." (Order: Homoptera.)

Cocoon: See *Pupa.*

Cricket: Insect with hind legs adapted for leaping and, in males, saw-like edges on the forewings that produce chirping sounds when rubbed together. About 2,400 species are known. (Family: Gryllidae.)

Dragonfly: Large, predatory, insect with long wings held out to the side of the body. The immature stages, or nymphs, are aquatic. Fast fliers, some dragonflies can fly 60 mph. With huge compound eyes that provide almost 360-degree vision, dragonflies arguably have the best eyesight of all insects. Eighty percent of dragonflies' brains are given over to sight; they have special eye-cleaning brushes on their front legs. More than 5,000 species are known. (Order: Odonata.)

Entomology: The study of insects.

Entomophagy: Eating insects. From the Greek *éntoma* (insects) and *phágein* (eating).

Exoskeleton: Hard external covering providing protection and support for an organism.

Fly: Technically, a small, soft-bodied insect with two wings (most insects have four—over time, the hind wings of flies evolved into stabilizing knobs called "halteres"). True flies include gnats, midges, black flies, house flies, and mosquitoes. About 90,000 species are known. Colloquially, any small winged insect. (Order: Diptera.)

Gall: An area of abnormal plant growth produced in response to "foreign agents," such as bacteria, fungi, viruses, or insects. Among the gallmaking insects are thrips, aphids, beetles (notably weevils), moths, wasps, and flies.

Grasshopper: Leaping insect with short antennae and eardrumlike hearing organs at the base of the abdomen or above the front feet. Many species can produce a buzzing sound either by rubbing together their front wings or toothlike ridges on the hind femurs against ridges on the front wings. (Families: Tettigoniidae and Acrididae [formerly Locustidae], Order Orthoptera.)

Grub: Thick-bodied, wormlike larva of some insects. See *Worm, Larva.*

Honey: Sweet, viscous liquid produced from flower nectar in the honey sacs of bees. After eating nectar, bees reduce its water content by exposing it to evaporation on their tongues; they also mix the nectar with enzymes to change its sugar content. After the transformation is complete, honey is regurgitated and stored in honeycomb, a double layer of hexagonal cells made of beeswax secreted by worker bees. Honey is made by, among other species, social bees in the genus *Apis*, especially the domesticated bee *A. mellifera.* See *Bee.*

Honeydew: Sugary excretory product of aphids and scale insects and other sucking insects that ingest the sugar-rich but protein-poor dilute sap of plants. In several ant species, foraging ants collect honeydew and bring it back to their nest, where special worker ants called "repletes" store honeydew in their abdomens, which distend to many times normal size. After stroking by worker ants, the replete regurgitates honeydew. Ants of at least six genera evolved this food-storage method independently. See *Ant.*

Insect: Evolutionarily the most successful animal group, the class Insecta consists of at least a million species, and perhaps as many as 30 million. Insects include beetles, butterflies, moths, ants, bees, and flies. Insects are characerized by three major body regions—head, thorax, and abdomen—and three pairs of walking legs.

Larva: Immature, wingless, often wormlike juvenile stage of an insect that undergoes complete metamorphosis. Sometimes called, imprecisely, "worms" or "grubs," insect larvae often look nothing like adults, caterpillars and butterflies being an example. See *Metamorphosis.*

Locust: Specifically, several species of short-horned grasshopper that often increase greatly in number and become destructive swarms. In North America the terms "locust" and "grasshopper" are used interchangeably for species in the family Acrididae. Cicadas are also loosely called "locusts"; the "17-year locust" is actually a cicada. See *Cicada, Grasshopper.*

Mealworm: Technically, the wormlike larva of the darkling beetle, in the genus *Tenebrio*, which infests granaries and bakeries. Generally, any larva of similar appearance that infests flour or meal.

Metamorphosis: Striking alteration in an organism, especially noted in insects, which undergo a sequence of regular changes called a "life-cycle." In the simplest insects, like silverfish and springtails, juveniles simply grow bigger, periodically shedding, or molting, their exoskeletons (ametabolous development). In more complex insects, such as grasshoppers, termites, and true bugs, the immature insect, or nymph, generally resembles the adult, but differs in coloration and proportion and in lacking fully formed wings. It gradually assumes adult form by molting and acquires wings in its final molt to adulthood (incomplete metamorphosis or hemimetabolous development). Beetles, butterflies, wasps, and true flies have wingless, often wormlike larvae, which do not resemble adults. They gather food to prepare for the inert pupal stage, in which wings and other adult features appear (complete metamorphosis, or holometabolous development). See *Larva, Nymph, Pupa.*

Moth: Nocturnal insect that resembles the butterfly but has a stouter body with hairlike or feathery antennae lacking knobs at the tip. Moths and butterflies have similar life-cycles. (Order: Lepidoptera.)

Nymph: Immature stage of an insect that undergoes incomplete metamorphosis. See *Metamorphosis.*

Parasite: An organism that grows, feeds on, and is sheltered by another organism while contributing nothing to its survival. Parasites aren't necessarily sheltered by hosts. Adult female mosquitoes, e.g., are ectoparasites; so are fleas and bed bugs. See *Symbiosis.*

Pheromone: Secreted chemical that influences the behavior or development of other animals of the same species.

Pupa: Nonfeeding stage between larva and adult in complete metamorphosis, during which the larva often encloses itself within a protective cocoon (a shroud of silk or other fiber produced by the larva) or chrysalis (a hard shell that hangs by a silklike thread). See *Metamorphosis.*

Sago: Food starch prepared from material in the trunks of palms, especially *Metroxylon rumphii* and *Metroxylon sagu* in Indonesia.

Scorpion: Nocturnal arachnid with a segmented body, an erectile tail tipped with a venomous sting, and like other arachnids, six pairs of appendages. The small first pair (chelicerae) tear apart prey. The second pair (pedipalps) have big, clawlike pincers that hold prey while the chelicerae work. All arachnids have segmented bodies. (Order: Scorpionida.)

Silk: Fine, lustrous fiber produced by insects for cocoons and webs. Commercial silk generally is produced by larvae of Asian moth species in the genus *Bombyx.*

Soldier: Caste of any social insect (ants, termites, even bees) that defends the colony from invaders. Wingless and usually lacking eyes, soldier termites protect the colony using their enlarged jaws or by emitting a sticky liquid that entangles enemies.

Spider: Predatory arachnids with two body regions joined by a waistlike constriction. Spiders are also characterized by spinnerets, abdominal organs used for spinning silk. Silk is used for many purposes, including for construction of webs to entrap prey, which are usually killed by venom injected via the fangs. About 34,000 species are known. (Order: Araneida.)

Stink bug: Broad, oval-shaped insect that emits a foul secretion, making the insect unappetizing to predators. More than 5,000 species are known. (Family: Pentatomidae.)

Stridulation: Shrill, grating sounds created by rubbing together specialized body surfaces. See *Cricket, Grasshopper.*

Symbiosis: Arrangement between two species in which at least one is dependent on the other. In mutualism, both species benefit from the arrangement. An example is the protozoans in the stomachs of many termites (though not all), which digest their wood. Because the protozoans cannot obtain wood themselves and the termites cannot digest wood unaided, neither can survive without the other. In commensalism, one species obtains food or other benefits from the other without affecting it, as in the pilot fishes that feed on leftovers from sharks. Parasitism is also a form of symbiosis. See *Parasite.*

Tarantula: Big, hairy, slow-moving spider, with jaws that move parallel to the body, rather than laterally (as in most other spiders). Most tarantulas are not seriously poisonous to humans but may inflict a painful bite if provoked. (Family: Theraphosidae.)

Taro: Widely cultivated tropical Asian plant (*Colocasia esculenta*) with broad leaves and a starchy, edible tuber.

Termite: Social, nest-dwelling insect with two pairs of wings that are nearly identical in size and shape. Termite colonies typically contain three castes: reproductives, workers, and soldiers. Nests are underground or in wood. Unlike ants and bees, termite populations contain both sexes in equal numbers. As with ants and bees, workers and soldiers are incapable of reproducing. Usually, each colony has one pair of reproductives, a king and a much larger queen. Workers, the most populous caste, and soldiers are generally blind. About 2,000 species are known. Also called "white ants," although they are not ants. Not all termites eat wood. (Order: Isoptera.) See *Soldier, Symbiosis, Worker.*

Urticating hairs: Detachable, stinging defensive hairs on some caterpillars and tarantulas. The term comes from the latin word for "nettle."

Warning coloration: Bold color patterns characteristic of a poisonous or unpalatable organism. Acts as a warning to predators.

Wasp: Stinging insect with narrow "waist" between the abdomen and thorax. More than 20,000 species are known, of which only a thousand are social. Like adult bees, adult wasps feed primarily on nectar; unlike bees, wasp larvae eat insects provided, in solitary or non-social species, by the female parent. (Suborder: Apocrita.)

Worker: Sterile member of a social insect colony responsible for brood care, nest construction and maintenance, foraging, and sometimes defense. Termite workers can be male or female, but ant or bee workers are always female.

Worm: Taxonomically ill-defined word used to refer to invertebrates with soft, slender, elongated bodies, often without obvious appendages. Many worm species are only distantly related to each other. Commonly but inaccurately, the term is often applied to centipedes, millipedes, some larval insects, and even to some vertebrates, such as the blindworm (*Anguis fragilis*), a limbless, snakelike lizard.

Glossary Sources

Berenbaum, May R. (1995), *Bugs in the System: Insects and Their Impact on Human Affairs.* Reading, Mass.: Addison Wesley.

Borer, D.F., R.E. White (1996). *A Field Guide to the Insects.* Boston: Houghton-Mifflin.

Encyclopædia Britannica. Chicago.

Hubbell, S. (1993). *Broadsides from the Other Orders: A Book of Bugs.* New York: Random House.

Wilson, E. O., B. Holldober (1990). *The Ants.* Cambridge Mass.: Belknap Press.

Sources and Resources

Recommended Reading

Amazing Bugs, Miranda Macquitty. New York: DK Publishing, 1996.

The Ants, E. O. Wilson and Bert Hölldobler. Cambridge, Mass.: Belknap Press, 1990.

The Book of Spiders and Scorpions, Rod Preston-Mafham. New York: Barnes & Noble, 1996.

Broadsides from the Other Orders: A Book of Bugs, Sue Hubbell. New York: Random House, 1993.

Bugs and Beetles, Ken Preston-Mafham. Edison, N.J.: Chartwell Books, 1997.

Bugs in the System: Insects and Their Impact on Human Affairs, May R. Berenbaum. Reading, Mass.: Addison-Wesley, 1995.

Butterflies in My Stomach, Ronald Taylor. Santa Barbara, Calif.: Woodbridge Press Publishing Co., 1975.

Caterpillars, Bugs and Butterflies: Take-Along Guide, Mel Boring. Minocqua, Wis.: NorthWord Press, 1996.

Creepy Crawly Cuisine, Julieta Ramos-Elorduy. Rochester, Vt.: Park Street Press, 1998.

Dining with Headhunters, Richard Sterling. Freedom, Calif.: The Crossing Press, 1995.

Eat Not This Flesh, Frederick J. Simoons. Madison, Wis.: The Univ. of Wisconsin Press, 1994.

Ecology of Food and Nutrition: An International Journal, V. 36, No. 2-4, Harriet V. Kuhnlein, Peter L. Pellett, Christine S. Wilson, *et.al.* Amsterdam: Gordon and Breach, 1997.

The Encyclopedia of Insects, Christopher O'Toole. New York: Facts on File, 1995.

Entertaining with Insects, or: The Original Guide to Insect Cookery, Ronald Taylor and Barbara Carter. Santa Barbara, Calif.: Salutek Publishing Co., 1996.

A Faber Book of Food. Colin Spencer and Claire Clifton. London: Faber and Faber, 1993.

Food in China: A Cultural and Historical Inquiry, Frederick J. Simoons. Boca Raton, Fla.: CRC Press, 1991.

Food in History, Reay Tannahill. New York: Crown Paperbacks, 1993.

The Food Insects Newsletter, Florence V. Dunkel, Ed., Dept. of Entomol., Montana State University, 324 Leon Johnson Hall, Bozeman, MT. 59717-0302, e-mail: ueyfd@montana.edu.

History of Food, Maguelonne Toussaint-Samat. Cambridge, Mass.: Blackwell Publishers, 1994.

Incredible Bugs, Rick Imes: New York: Barnes & Noble, 1997.

Insect Potpourri: Adventures in Entomology, Jean Adams. Gainesville, Fla.: Sandhill Crane Press, 1992.

The Insect Societies, Edward O. Wilson. Cambridge, Mass.: Harvard University Press, 1974.

Material World: A Global Family Portrait, Peter J. Menzel. San Francisco: Sierra Club Books, 1994.

Ninety-Nine Gnats, Nits, and Nibblers, May R. Berenbaum, Chicago: University of Illinois Press, 1989.

Ninety-Nine More Maggots, Mites, and Munchers, May R. Berenbaum, Chicago: University of Illinois Press, 1993.

On Food and Cooking: The Science and Lore of the Kitchen, Harold McGee. New York: Macmillian Publishing, 1984.

The Practical Entomologist, Rick Imes. New York: Simon & Schuster, 1995.

That Gunk on Your Car: A Unique Guide to Insects of North America, Mark Hostetler. Berkeley: Ten Speed Press, 1996.

Uniquely Australian: The Beginnings of an Australian Bushfood Cuisine, Vic Cherikoff. Sydney: Bush Tucker Supply Party Ltd, 1992. www.bushtucker.com.au.

Women in the Material World, Faith D'Aluisio and Peter Menzel. San Francisco: Sierra Club Books, 1996.

Yanomamo, Napoleon A. Chagnon. Ft. Worth, Tex.: Harcourt Brace College Publishers, 1997.

Places to See Bugs

The Arachnid Exhibit at the Louisville Zoo—1100 Trevilian Way, Louisville, Kentucky 40213. Admission fee. 502-459-2181.

"Bug" World at Woodland Park Zoo—5500 Phinney Avenue N., Seattle 98103-5897. Admission fee. 206-684-4800.

Butterfly Pavilion and Insect Center—6252 W. 104th Avenue, Westminster, Colorado 80020. Admission fee. 303-469-5441.

Butterfly World—Tradewinds Park S., 3600 W. Sample Road, Coconut Creek, Florida 33073. Admission Fee. 954-977-4400.

Cincinnati Zoo's World of Insects—3400 Vine Street, Cincinnati, Ohio 45220. Admission fee. 800-94-HIPPO.

Cockrell Butterfly Center and Insect Zoo, Houston Museum of Natural Science—One Hermann Circle Drive, Houston, Texas 77030-1799. Admission fee. 713-639-IMAX.

The Insectarium at Steve's Bug-Off Exterminating Company—8046 Frankford Avenue, Philadelphia, Pennsylvania 19136. Admission fee. 215-338-3000.

Moody Gardens—One Hope Blvd., Galveston, Texas 77554. Admission fee. 800-582-4673.

National Zoo's Invertebrate Exhibit and Pollinarium—3001 Connecticut Avenue N.W., Washington, D.C. 20008. Free admission. 202-673-4800.

Niagara Parks Butterfly Conservatory—2565 N. Niagara Parkway, Niagara Falls, Ontario L2E 6T2, Canada. Admission fee. 905-356-8119.

Otto Orkin Insect Zoo at the National Museum of Natural History, Smithsonian Institution—10th Street & Constitution N.W., Washington, D.C. 20560. Free admission. 202-357-2700.

Ralph M. Parsons Insect Zoo, Natural History Museum of Los Angeles County—900 Exposition Blvd., Los Angeles, California 90007. Admission fee. 213 763 3558.

San Francisco Zoo's Insect Zoo—1 Zoo Road, San Francisco, California 94132-1098. Admission fee. 415-753-7080.

Bug Fact Sources

Page 28: Cherry, R.H. (1991). *American Entomologist:* 37:8-13. Lanham, Md.: Entomological Society of America.

Page 33: Mitsuhashi, J. (1997). *Ecology of Food and Nutrition,* 36:2-4, 189-190. Amsterdam: Gordon and Breach Science Publ.

Page 43, col. 1: Ichikawa, N. (1995). *Journal of Insect Behavior,* 8:181-188. *The Food Insects Newsletter* (1990). 3(2):6. Bozeman, Mont.

Page 43, col. 3 and Page 101: Berenbaum, May R., Univ. of Illinois at Urbana-Champaign.

Page 51: *Encyclopædia Britannica* Micropædia. (1995). Vol. 3, pp. 735, Chicago. Imes, R. (1992). *The Practical Entomologist* (pp. 79). New York, NY: Simon and Schuster.

Page 74: Baker, D. (1987). "Foreign Bodies of the Ears and Nose in Children," *Pediatric Emergency Care,* 3:67. Baltimore, Md.: Williams and Wilkins. O'Toole, K., P.M. Paris, and R.D. Stewart, (1985). "Removing Cockroaches from the Auditory Canal: A Controlled Trial," *The New England Journal of Medicine,* 312:1192. Waltham, Mass.: Massachusetts Medical Society.

Page 89: *The Food Insects Newsletter* (1993). 6(3):3. Chen, *et. al.* (1997). *Life Sciences,* 60:2349-2359. New York: Elsevier Science.

Page 98: Tu, A.T. (1984). "Insect Poisons, Allergens and Other Invertebrate Venoms," *Handbook of Natural Toxins,* Vol. 2. New York: Marcel Dekker.

Page 99: Scott Stockwell, Ph.D, Academy of Health Sciences, Ft. Sam Houston, Tex.

Page 116: *The Food Insects Newsletter* (1991). 4(1):6.

Page 141: Gaston, K.J., S.L. Chown, C.V. Styles (1997). "Changing Size and Changing Enemies: The Case of the Mopane Worm," *Acta Oecologia,* 18:21-26.

Page 149: Bukkens, S. (1997). *Ecology of Food and Nutrition.* 36:287-319. Amsterdam: Gordon and Breach Science Publ. Wilson, E.O. (1971). *The Insect Societies,* Cambridge, Mass.: Harvard Univ. Press. *The Food Insects Newsletter* (1996) 9(1):7.

Page 160: Hanks, L.M. *et al.,* (1996). *Journal of Insect Behavior,* 9:383-393. New York: Plenum.

Page 174: Marshall, S.D., Thoms, E.M., and Uetz, G.W. (1995). *Journal of Zoology,* 235:587-595.

Page 178: *The Food Insects Newsletter* (1989). 2(2):3.

Acknowledgments

This book is dedicated to Gene DeFoliart, who taught us about entomophagy with good humor and a grain of salt.

Acknowledgments

Our deep gratitude to the following people:

Staff members Sheila DS Foraker and Jean Pihl; Charles C. Mann; David Griffin; Tim Cahill, May R. Berenbaum, vegetarian entomologist, for her unbiased expertise; Gene DeFoliart; Chelsea Vaughn, our editor at Ten Speed Press, and everyone at TSP; Dick Lemon, Esq.; Akilah Jaye; Todd Rogers; and Josh D'Aluisio-Guerrieri.

Sam Hoffman and everyone at The New Lab; Gary Young, Ken Lien at Kodak Digital Imaging; Roni Kendall, Xanté; Sandra Bukkens and Mario Paoletti; Marc Weiser, Xerox Parc; Michael Hawley and Rob Silvers, MIT Media Lab; Monica Cipnic, *Popular Photography;* David Friend; Ruth Eichorn, *GEO;* Ray Kinoshita; Robin and Dawn D'Aluisio; Florence Dunkel; Laura Hunt; Liv Mills-Carlisle; Jack and Evan Menzel; Adam D'Aluisio-Guerricri; Newell and Emelia K. Mann; and Kyle Griffin.

Australia: Vic Cherikoff, Bush Tucker Supply Party Ltd.; Merle Scarce, Institute for Aboriginal Development. **Botswana:** Ben van der Waal. **Cambodia:** Bernard Krisher, and The Cambodia Daily; Chung "Tana" Chotana; Yeang "Kohn" Sokhon. **China:** Geographic Expeditions, San Francisco; Shen Mao Mao; Josh D'Aluisio-Guerrieri; Scott Stockwell; Dr. Wen-jun Hseih; Hong Kong Fishing Village Restaurant, Guangzhou; Professor Xiaoming Chen and Professor Fang; Mr. Lee, Kunming; Gu Xin Scorpion Farmland, Luoyang; Professor Lu A Ping, Zhongshan University, Guangzhou; Han Jin and Hou Song Feng; Ru Yang Boda Scorpion Breeding Company, Ltd. **Indonesia:** Dennis Kooren, Geographical Connections; John Cutts. *In Jayapura:* Sam Maturbongs; Andre Liem. *In The Asmat:* Marius Revideso, our translator; Luki Kanuga; Rony Kogoya; Kondradus Kamau; Bishop Alphonse Sowedo; Fathers O'Brien, Antoine, and Jim; Father Virgil Peterman; Father Trenk. *In Komor:* Plipus, Adrienne, and the entire village! *In Sawa:* Father Vince Cole; Victor Aunam; Pascales Buap; Rufus Siti; Victor Pardamok; Rufina Dochan; Udelia Toronam; Udelia Fumyap; William Pupice; Vincent Toman; Rufus Ausay; Sabinus Weganap; Adriana Toranam; Dorfinis Aumbus; Alfonse Damar; Bertila Semap; Fincenu Tobepar; Januarius Awop; Ava Tofifirar; Pius Tenacam; Sabinus Barap; and Pius Sermar. *In Baliem Valley:* Thony Ngamel; Musa Wili; Buzz Maxie; Isaak Wantikbo. *In Soroba village:* Edo Himan; Augus Wilil; Aloka Wilil; Marta Himan; Martinus Himan; Yorim Surabut; Oloma Kosay; Silumatek Wilil; and Wori Uluwa. *Bali:* I Wayan "Darta" Sudarta; Oka; I Wayan Taweng; I Wayan Darsana; Imade Griyawan; Eka Mahardika; Imade Ari Bawa; Selemet Susila; Sri Yani; Mardana; Ari Widyastuti. **Japan:** Uniphoto Press International. **Mexico:** Julieta Ramos-Elorduy; Anna Sever; Tino; Carol Starling; Cresciana Rodríguez Nieves and her husband; Pedro Valle Vega, Ph.D.; Patrick Chovela; José "Paco" Francisco Sarnane and Alejandro, Jr.; and Ignacio Martínez García. **Thailand:** The Kuenkaew family; Professor Prachaval Sukumalanand; Aikkarach Kettawan; Suwanna Tantayanusorn. **Peru:** Steven King, Shaman Pharmaceuticals; Nilda Callañaupa and her husband; Ángel Callañaupa; Raúl Jaimes, our transla-tor; Wilson Jaimes; Selso Jaimes. *In Chicón:* Nicasio Huaman Callañaupa; Asensia Huaman; Carmen; Irene; and Eldelberto. *In Chinchapujio:* Bernadina Ochoa; Paulina Wilca-Ochoa; Ana Ruth; José; Victor; and Nilson. Salvador F. Ticona Ramos; Inez; Irmalinda; Gloaldo; Dennis; and Aldair. Crisptín Carasco; Ironima Soto Varez; and Américo Carasco Soto. *In Koribeni:* Daniel Piña Real and Elena; Marleni; Ramiro; Sylvia; Karen; and Daniel Jr. *In Yaneriato:* Solomon Mintiani; Lucía Tenteyo; Leonel Mintiani; Mari Luz; Delia; Rosa; Edmundo; Mintiani; Margarita Yoveni; Máximo Katiga; Florinda Kashiari. **South Africa:** Ben and Willie van der Waal; Tshifhiwa Eric Munzhedzi; Peggy; Eric's father and mother; The Termite Club: Mrs. Humbulani; Joyce Netshipise; Violet Nemagovhhani; Patricia Maphangule; Florence Nemagovhani; Nomsa Mpondi. Thanks also to Auhatarkati Netshipise for her dried *ntwa* termites. Jerry Madayhu; Sylvia; Duncan; and Itani. **Uganda:** Andrew Galiwango; Olivia Maganyi, our excellent translator; Tina Vivera; Susan Nelson, Shaman Pharmaceuticals. The villages of Bweyogerere and Katalemwa in Mpigi District and Mawejje Dick's family. **United States:** Larry Peterman and Dena Cagliero, HotLix; Dale, Dorothy, Russ, and John Bassett, Bassett's Cricket Ranch; Safiya Carter-Thompson and her mother Linda; Leslie Saul and Norman Gershenz, Center for Ecosystem Survival at San Francisco State University. **Venezuela:** Carlos Peña; Javier Mesa; Mark Moffit, Ph.D.; Juan Carlos Ramírez; and Hugo Cerda, Ph.D.

A Material World Book

Ten Speed Press
P. O. Box 7123 • Berkeley, California 94707
www.tenspeed.com

Distributed in Australia by Simon & Schuster Australia, in Canada by Ten Speed Press Canada, in New Zealand by Tandem Press, in South Africa by Real Books, in the United Kingdom and Europe by Airlift Books, and in Singapore and Malaysia by Berkeley Books.

Library of Congress Cataloging-in-Publication Data

Menzel, Peter, 1948–

    Man eating bugs: the art and science of eating insects / by Peter Menzel and Faith D'Aluisio.
    p. cm.
    "A Material world book."
    ISBN 1-58008-022-7
    ISBN 1-58008-051-0
    1. Entomophagy. 2. Edible insects. 3. Food habits.
I. D'Aluisio, Faith, 1957– . II. Title
GN409.5.M46 1998
394.1--dc21

            98-4411
            CIP

First printing, 1998
Printed in Hong Kong
1 2 3 4 5 6 7 8 9 — 02 01 00 99 98

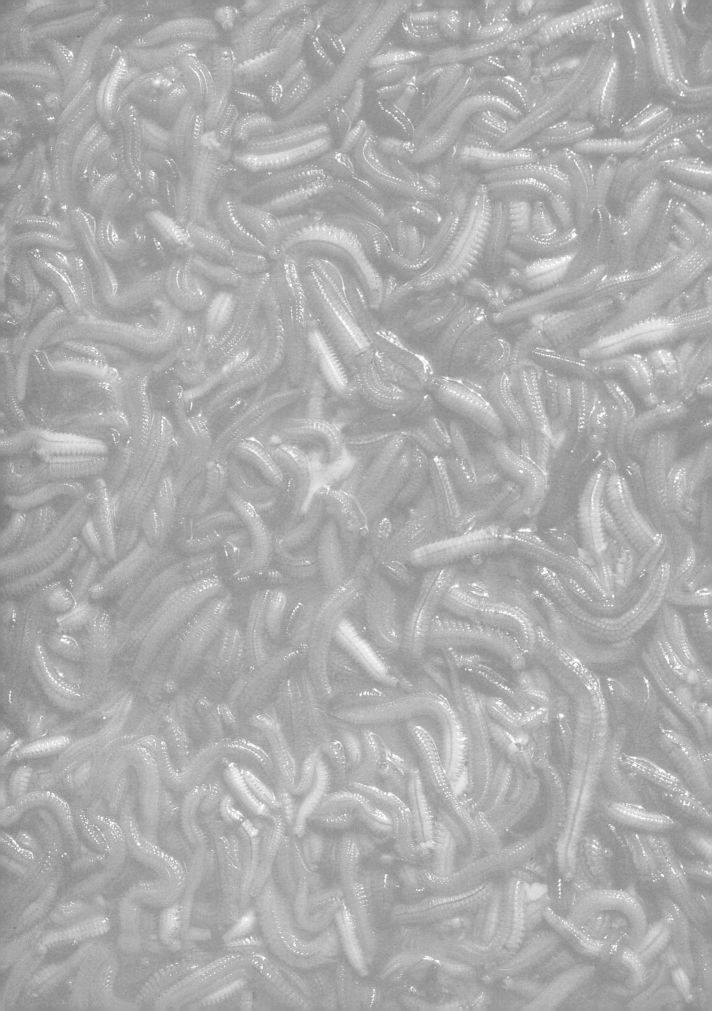